I0397504

PSYCHOTHERAPY, CONCEPTS OF TREATMENT

ISBN: 978-1-291-50178-0

Andreas Sofroniou   2013 ©   Copyright

Andreas Sofroniou   2013 ©   Copyright

PSYCHOTHERAPY, CONCEPTS OF TREATMENT

ISBN: 978-1-291-50178-0

CONTENTS   *Page*

1. PSYCHOTHERAPY EXPLAINED

The simplistic explanation of psychotherapy is that it deals with the treatment of disorders of emotion or personality by psychological methods. Formerly, this was the treatment of disease by psychic or hypnotic influence. As the profession of psychotherapy is now established, the treatment of emotional or behavioural problems by psychological means, often in one-to-one interviews or small groups, is now the accepted norm.

The modern type of psychotherapy started with Sigmund Freud who devised the first systematic approach, initially discussing patients' problems with them, but later allowing them to do most of the talking in a procedure called free association of ideas. This has been the model for subsequent psychotherapies.

Modern psychoanalysis and cognitive therapies associated with theories such as learned helplessness concentrate on the patient's beliefs. Other therapies, such as those within humanistic psychology, attend to the patient's emotional state or sensitivity. The distinction, however, is not clear-cut, as all these therapies involve intense exploration of the patient's conflicts, and most rely on the emotion generated in therapy as a force in the patient's recovery. In contrast, behaviour therapies derive from the view that neurosis is a matter of maladaptive conditioning and concentrate on modifying patients' behaviour.

There are arguments about the effectiveness of psychotherapies, but it is generally agreed that success depends on a secure, confiding relationship between the therapist and patient and on a shared confidence in the capacity of the therapist and his or her theory to explain and eliminate the problem.

It will be of more assistance if at this point, the Freudian concept of therapy is explained further. Freud, Sigmund (1856-1939), was an Austrian psychologist and psychotherapist who pioneered psychoanalysis. Following a medical training, in which he specialized in neurology, he turned to the study of hysteria. *Studies on Hysteria* (1895), written with Josef Breuer (1842-1925), established the framework of psychoanalytic theories about neurosis: that the symptoms result from (and have a symbolic relation to) an emotional trauma which the patient has 'forgotten'. The memory, however, acts in the unconscious to disrupt the patient's thoughts and feelings. To gain access to unconscious material the therapist uses hypnosis, dream analysis, or free association.

Freud published *The Interpretation of Dreams* in 1900. He noticed the importance of sexual themes in this material and argued that sexual 'memories' were really childhood fantasies which represented infantile sexual desires. This thesis of childhood sexuality, laid out in *Three Essays on the Theory of Sexuality* (1905), became central to Freud's thinking about normal and abnormal psychosexual development. His theories came from clinical observations about which he wrote case histories; some of which, however, were subsequently shown to have been self-censored. There are many sources of distortion in such data and Freud's preoccupation with sexual explanations sometimes led him to ignore more obvious possibilities.

Freudian theory enjoyed great popularity in the 1940s and 1950s, when academics such as the social learning theorists at Yale University in the USA attempted to test his ideas by experiment, but they were rarely confirmed. Freud and his followers developed such key concepts as displacement, identification, projection, regression, repression, and sublimation and defence mechanism. His ideas remain controversial and influential not only in the treatment of psychiatric illness, but in the arts and social sciences. Later psychoanalytic theorists who followed Freud are known as neo-Freudian.

In recent times psychoanalysis was well recognised as a theory of and therapy for the mental disorders known as neuroses, and a general theory of personality and emotional development constructed almost entirely by Freud. The therapy is one-to-one, over an extended period, and investigates the interaction between the conscious and, by free association of ideas, the unconscious mind, bringing to light repressed fears and conflicts. The theory has been of enormous influence in 20th-century thinking and culture. It gives a central role to the drives of instinct and the way that socialization may pervert such drives through too much indulgence or control.

Psychoanalysis stresses that instincts and emotions may remain, unacknowledged, in the unconscious and profoundly affect thought and behaviour. Freud believed that instincts from childhood onwards revolve round physical gratification and are broadly sexual. Subsequently, however, he suggested that we have destructive as well as sexual instincts.

The Austrian-British psychoanalyst Melanie Klein (1882-1960) took this idea of *Thanatos* (death) and *Eros* (love) much further; her work with children is probably the most important contribution to psychoanalysis after Freud. At this period the anthropologist Malinowski demonstrated that the importance of sexual instincts in personality may not be a universal feature of human development.

In *The Ego and the Id* (1923), Freud proposed a tripartite division of the personality into *ego* ('I'), *id* ('it'), and *super-ego*. The id is unconscious and contains primitive emotions and drives. The super-ego contains ideals and moral values. The ego steers between the two, trying to reconcile their demands and the constraints of the real world. It is the seat of consciousness, but its defence mechanisms are not conscious processes. Later neo-Freudian theorists placed more stress on development of the ego and less on the unconscious and sexual motivation (Jung rejected the centrality of the latter).

The theory of psychoanalysis, which is arguably the most inclusive in psychology, has fundamental conceptual weaknesses. Many of its claims are impossible to test by experiment, or, where tested, have not been confirmed. Psychoanalytical therapy seems less effective than some more economic forms of psychotherapy, behavioural therapy, and other treatments offered by psychiatry.

2. TYPES OF PSYCHOTHERAPY

HUMANISTIC PSYCHOLOGY

This is an approach to psychological study and therapy which reacts against mainstream psychology's apparent willingness to treat humans as objects of investigation and manipulation.

Humanistic psychology emphasizes co-operation, empathy, and mutual respect between psychologist and patient ('client'). Its philosophical ancestry is traceable through existentialism to Rousseau. Key concepts for most humanistic psychologists are the 'real self' and 'self-actualization', as expounded by the Scottish psychiatrist R(onald) D(avid) Laing (1927-90) and the US psychotherapist Rogers.

Humanistic psychologists encourage self-expression, self-discovery, and 'personal growth' through insight and emotional experience. There is no clear set of doctrines or single therapeutic approach. Techniques are drawn from many traditions, and include forms of exercise to foster bodily awareness and reduce tensions, and also self-disclosure or role-playing, as in psychodrama and group therapy.

Humanistic psychology offers an appropriate attitude for psychotherapy, but may have limited use in challenging the experimental approach of academic psychology because the two are based on different premises.

BEHAVIOUR THERAPY

This school of ideas is also known as behaviour modification, by applying clinical treatment to change maladaptive patterns of behaviour using techniques which derive from experiments on learning. It is based on the view that neurotic behaviour is learnt in the same way as normal behaviour and may be eliminated by a manipulation of surroundings which allows the subject to learn new and better responses.

It developed in the 1950s from Watson and Skinner's ideas about therapeutic applications of conditioning. The classic text was Joseph Wolpe's *Psychotherapy by Reciprocal Inhibition* (1958). He described how he had made cats fearful and then cured them by inducing them to eat closer and closer to the source of their fear.

Behaviour therapy has become popular, in the form of desensitization, for treating phobias and obsessions, and has been used with some success for sexual problems and alcohol and drug abuse.

It has also been used to try to 'modify' anti-social activities in classrooms. Generally, however, the co-operation of the patient seems necessary. Behaviour therapy does not try to rearrange patients' understanding of themselves or their world. It may therefore appear somewhat mechanistic. More recent approaches focus on patients' beliefs about what they are able to do.

CONDITIONING

In psychology, this is a change in behaviour due to association between events. It was the basis of learning theories which dominated academic psychology from World War I to about 1960. Conditioning is usually divided into two kinds: classical or Pavlovian; and operant or instrumental. Both involve the pairing of an event with 'reinforcement', which may be 'positive' (rewards of food, drink, or sex) or 'negative' (punishment such as electric shock).

In classical conditioning, which was discovered by Pavlov, a light or sound is paired with a natural reinforcement. The response which was initially produced by the reinforcement becomes 'conditioned' so that it occurs to the light or sound even when no reinforcement is given. This is therefore a matter of learning an association between two stimuli (the reinforcement and the light or sound) and is referred to as S-S conditioning.

Operant conditioning follows the US psychologist Edward Thorndike's (1874-1949) 'law of effect' (1911): that responses become more frequent if followed by satisfying consequences but less frequent if followed by aversive consequences. Skinner showed that a rat which is rewarded when it 'operates on' its environment by pressing a lever will increase its number of lever-presses. It is therefore associating the stimulus (reinforcement) with its own behaviour (response). This is referred to as S-R conditioning.

Psychologists dispute whether these two kinds of conditioning do really differ from each other. Most conditioning experiments have been done with animals. It is very doubtful whether all animal, let alone human, learning is due to conditioning. However, Skinner pointed out that it plays a role, for instance in the unwitting encouragement of misbehaviour when parents reward a child by attending to misbehaviour but ignoring good behaviour. In 1920 Watson showed that fears can be conditioned and thereby laid the foundations for behaviour therapy treatments for phobia.

CLINICAL PSYCHOLOGY

It refers to the application of procedures derived from theory and research in psychology to the assessment and treatment of mental and physical disorders. The term was first used in the 19th century to refer to methods of assessing physical and mental handicap. Assessment of clinical conditions such as brain damage developed during the two World Wars.

After World War II clinical psychology also became important for procedures of rehabilitation and especially for the various psychotherapies.

Many of the latter derived initially from psychoanalysis, but the research on which learning theories were based gave rise to treatments which did not rely on drugs or make medical assumptions about abnormal behaviour being an 'illness'. These behaviour therapies have proved fairly successful in the treatment of phobias and some other disorders.

Clinical psychologists are now likely to be eclectic rather than advocates of particular theoretical or therapeutic models. Nevertheless, the view that many abnormalities are at least partly caused by experience ('faulty learning'), and can be improved by procedures akin to those of conditioning remains fundamental to most clinical psychologists.

PSYCHIATRY

Psychiatry is the branch of medicine concerned with the study and treatment of mental illness. The conditions coming within its scope are mainly those, in which there is no established damage to, or disease of, the brain (these are the province of neurology) and include psychoses such as schizophrenia; neuroses such as phobias; depression; dementia; and personality disorders.

Psychiatrists, who generally work with other professionals such as psychiatric nurses, clinical psychologists and social workers, offer varied treatments. These include drug treatment, electro-convulsive therapy (ECT) (for some forms of depression), behaviour therapy, family therapy, group therapy, psychotherapy, and individual counselling and support.

The practice of psychiatry is fraught with difficulties. The ethical permissibility of many treatments has been challenged on the grounds that they may have damaging side effects, be of questionable efficacy, and may pose a threat to patients unable, because of illness, to give informed consent. Another problem is that although only a small proportion of mentally ill people are a threat to themselves and an even smaller proportion a threat to others, dangerousness cannot accurately be predicted.

The claim by the Hungarian-American psychiatrist Thomas Szasz in *The Myth of Mental Illness* (1961) that psychiatrists were modern-day inquisitors labelling behaviour in order to control people's conduct, was one of the catalysts of a movement which has resulted in the USA and elsewhere in deinstitutionalization and in restrictions by law on the powers of psychiatrists to detain and treat the mentally ill.

In response, psychiatrists argue that mental illness, far from being a myth, is to be found in every society, is shown by research to be the result of the interaction of biological (probably inherited) predispositions and environmental factors, and is the cause of immense distress to sufferers and their families for which treatment, even if imperfect, is sought.

In 1977 the psychiatrists' international body, the World Psychiatric Association, promulgated the Declaration of Hawaii, a set of ethical guidelines to promote high standards and prevent the misuse of psychiatry, not least in countries where political dissenters are diagnosed as mentally ill and incarcerated.

COMPLEMENTARY MEDICINE

Alternative medicine includes a variety of forms of health care that fall outside the official health sector. Such health care provides an alternative to Western or allopathic medicine, which complementary practitioners believe treats symptoms and diseases rather than individuals in their complex physical, emotional, and environmental contexts.

Formalized traditional systems of medicine, such as Ayurvedic and Chinese medicine, or the practice of traditional healers in Africa, can be termed complementary, as well as newer therapies practised in the Western world, some of which are based on the theories or practices of traditional medicine.

Such therapies include acupuncture; homoeopathy; chiropractic and osteopathy (forms of manipulation); and herbal medicine. In the UK, the practice of these therapies requires training and registration. There are other diagnostic or therapeutic techniques that might be termed fringe medicine.

They include iridology (diagnosis from an examination of the eye); reflexology (treatment by massaging the foot in order to cure ailments in other parts of the body); and biofeedback (the use of monitoring equipment to help control involuntary processes, such as heart rate).

Many forms of paranormal healing also come into the category of complementary medicine. In the West there is a growing interest in such alternative therapies, although most conventional medical practitioners do not accept their claims.

Some doctors, however, practise holistic medicine, which is the combination of conventional medicine with forms of complementary medicine, self-help skills, and psychotherapy, in an attempt to treat the whole person and not merely his or her physical symptoms.

COUNSELLING

In psychology this is the guidance offered in the form of discussion rather than any specific type of therapy. Counsellors may have a quantity of specialized knowledge on particular problems, possible solutions, and the potential pitfalls of different solutions, but often they see their role as assisting those who consult them to find their own solutions.

Counselling may be available to help people adapt after physical injury, traumatic shock, when a family has been incapacitated by mental or physical injury, when a marriage is unhappy, when a child has been abused, or when career advice is needed. It may be offered by psychologists, doctors, social workers, or trained lay people.

In marriage guidance, the counsellor to some extent acts as an arbitrator, trying to help the couple reach compromises which might meet some of the demands of both parties. In career guidance, counsellors may administer various psychological tests to assess the clients' capacities and interests.

Counselling should be distinguished from psychotherapy, although there is some overlap. It is less intensive and sometimes more directive and its clients may have a much wider range of concerns for which they require guidance.

HYPNOTHERAPY

Hypnosis, hypnotherapy, or tonic immobility is a condition which refers to two slightly different phenomena, one in animals and one in humans.

Animal hypnosis is a natural phenomenon whereby the animal becomes immobile, or paralysed, as a reaction to fear. This is also known as tonic immobility, similar to catatonic trance in humans, which is often caused by fear too.

Tonic immobility can be experimentally induced in animals with a fear of humans by physical restraint, and in domestic hens, the duration of tonic immobility seems to depend on the level of fear induced, for example by transportation. In the wild, it is a last resort when an animal is attacked; it serves to reduce its chances of being killed and eaten; most predators respond to the movement of prey and will not attack what is apparently dead.

Human hypnosis is a trance-like condition involving apparent alteration in consciousness and memory, during which the hypnotized subject becomes susceptible to the hypnotist's suggestions. These may produce anaesthesia (absence of sensation), analgesia (insensitivity to pain), paralysis, or apparent regression to childhood functioning.

Post-hypnotic suggestions made during hypnosis influence the subject after the trance has ended and may induce the subject to forget what has occurred during hypnosis (post-hypnotic amnesia).

Hypnosis is a contentious issue. Some believe it is simply a form of role-playing in which subjects voluntarily act in a way they believe typical of those 'hypnotized'. They point to evidence that there is no physiological sign, for example in brainwave patterns, of any change in consciousness.

Others believe that hypnosis is a special state of consciousness, pointing out that similar trance-like states, often resulting from repetitive stimuli such as drumming, have been reported since ancient times and in all cultures.

Hypnotic suggestibility differs from person to person. A tendency to vivid daydreams and willingness to be hypnotized characterize those who are susceptible. Hypnosis has been used in dentistry, obstetrics, and psychotherapy.

In all cases the patient's belief in its efficacy seems crucial. It was the study of hypnosis which led Freud to explore the unconscious mind and develop his theory of psychoanalysis.

PERSONALITY ASSESSMENT

Although assessments and psychometrics have been used for tens of years the reliability and validity of assessment methods are still disputed by psychotherapists.

Assessment, whether it is carried out with interviews, behavioural observations, physiological measures, or tests, is intended to permit the evaluator to make meaningful, valid, and reliable statements about individuals. What makes John Doe tick? What makes Jane Doe the unique individual that she is? Whether these questions can be answered depends upon the reliability and validity of the assessment methods used.

The fact that a test is intended to measure a particular attribute is in no way a guarantee that it really accomplishes this goal. Assessment techniques must themselves be assessed.

EVALUATION TECHNIQUES

Personality instruments measure samples of behaviour. Their evaluation involves primarily the determination of reliability and validity. Reliability often refers to consistency of scores obtained by the same persons when retested. Validity provides a check on how well the test fulfils its function.

 The determination of validity usually requires independent, external criteria of whatever the test is designed to measure. An objective of research in personality measurement is to delineate the conditions under which the methods do or do not make trustworthy descriptive and predictive contributions.

One approach to this problem is to compare groups of people known through careful observation to differ in a particular way. It is helpful to consider, for example, whether the assessments and measurements discriminate significantly between those who show progress in psychotherapy and those who do not, whether they distinguish between law violators of record and apparent non-violators. Experimental investigations that systematically vary the conditions under which subjects perform also make contributions.

Although much progress has been made in efforts to measure personality, all available instruments and methods have defects and limitations that must be borne in mind when using them; responses to tests or interview questions, for example, often are easily controlled or manipulated by the subject and thus are readily "fakeable."

Some tests, while useful as group screening devices, exhibit only limited predictive value in individual cases, yielding frequent (sometimes tragic) errors.

These caveats are especially poignant when significant decisions about people are made on the basis of their personality measures. Institutionalization or discharge, and hiring or firing, are weighty personal matters and can wreak great injustice when based on faulty assessment.

In addition, many personality assessment techniques require the probing of private areas of the individual's thought and action. Those who seek to measure personality for descriptive and predictive reasons must concern themselves with the ethical and legal implications of their work.

A major methodological stumbling block in the way of establishing the validity of any method of personality measurement is that there always is an element of subjective judgment in selecting or formulating criteria against which measures may be validated. This is not so serious a problem when popular, socially valued, fairly obvious criteria are available that permit ready comparisons between such groups as convicted criminals and ostensible non-criminals, or psychiatric hospital patients and non-institutionalized individuals.

Many personality characteristics, however, cannot be validated in such directly observable ways (*e.g.*, inner, private experiences such as anxiety or depression). When such straightforward empirical validation of an untested measure hopefully designed to measure any personality attribute is not possible, efforts at establishing a less impressive kind of validity (so-called construct validity) may be pursued.

A construct is a theoretical statement concerning some underlying, unobservable aspect of an individual's characteristics or of his internal state. ("Intelligence," for example, is a construct; one cannot hold "it" in one's hand, or weigh "it," or put "it" in a bag, or even look at "it.") Constructs thus refer to private events inferred or imagined to contribute to the shaping of specific public events (observed behaviour).

The explanatory value of any construct has been considered by some theorists to represent its validity. Construct validity, therefore, refers to evidence that endorses the usefulness of a theoretical conception of personality. A test designed to measure an unobservable construct (such as "intelligence" or "need to achieve") is said to accrue construct validity if it usefully predicts the kinds of empirical criteria one would expect it to--*e.g.*, achievement in academic subjects.

The degree to which a measure of personality is empirically related to or predictive of any aspect of behaviour observed independently of that measure contributes to its validity in general. A most desirable step in establishing the usefulness of a measure is called cross-validation.

The mere fact that one research study yields positive evidence of validity is no guarantee that the measure will work as well the next time; indeed, often it does not. It is thus important to conduct additional, cross-validation studies to establish the stability of the results obtained in the first investigation.

Failure to cross-validate is viewed by most testing authorities as a serious omission in the validation process. Evidence for the validity of a full test should not be sought from the same sample of people that was used for the initial selection of individual test items.

Clearly this will tend to exaggerate the effect of traits that are unique to that particular sample of people and can lead to spuriously high (unrealistic) estimates of validity that will not be borne out when other people are studied.

Cross-validation studies permit assessment of the amount of "shrinkage" in empirical effectiveness when a new sample of subjects is employed. When evidence of validity holds up under cross-validation, confidence in the general usefulness of test norms and research findings is enhanced.

Establishment of reliability, validity, and cross-validation are major steps in determining the usefulness of any psychological test (including personality measures).

3. TREATMENT FOR DISORDERS

Psychotherapy is a form of treatment for psychological or emotional disorders in which a trained person establishes a relationship with one or several patients for the purpose of modifying or removing existing symptoms and promoting personality growth. Drugs may be used as adjuncts, but the healing influence is exerted primarily by words and actions that are believed by sufferer, therapist, and the group to which they both belong to have healing powers and that create an emotionally charged relationship between or among them.

Modern individual and group psychotherapeutic methods are used to treat all forms of suffering in which emotional factors play a part.

These include:

- behaviour disorders of children and adults;

- emotional reactions to the ordinary hardships or crises of life;

- psychoses, characterized by derangements of thinking and behaviour usually so severe as to require hospitalization;

- psychoneuroses, which are chronic disorders of personal functioning often accompanied by bodily symptoms of emotional strain;

- addictions;

- psychosomatic disorders, in which tissue damage is caused or aggravated by emotional components; and

- stress.

Psychotherapeutic principles are also emphasized in rehabilitation programs for the disabled and chronically ill.

Early treatment of mental illness was based on either a religio-magical or a naturalistic view of disease. The former, originating before recorded history, saw certain forms of personal suffering or of alienation from one's fellows as caused by an evil spirit that gained entrance into the sufferer. Treatment was based on participation in suitable rites under the guidance of a priest-physician, medicine man, or shaman. The naturalistic tradition viewed mental illness as a phenomenon that could be scientifically studied and treated. Treatment consisted of measures to promote bodily well-being and mental tranquillity.

Psychotherapy of non-hospitalized patients in the naturalistic tradition was not distinguishable from ordinary medical practice until the latter half of the 19th century. The emergence of psychotherapy as a specialized treatment probably is traceable to the late 18th century.

A dramatic demonstration by an Austrian mystic and physician, Franz Anton Mesmer, showed that many symptoms could be made to disappear by putting a patient into a trance. Mesmerism was the precursor of hypnotism, which became a widely used psychotherapeutic method.

Through it, Josef Breuer and Sigmund Freud together made the epochal observations on the relationship to later mental illness of emotionally charged, damaging experiences in childhood. From these discoveries grew the theory and practice of psychoanalysis, which, with its many modifications, immensely influenced the subsequent development of psychotherapy.

Modern psychotherapeutic methods for influencing patients directly include giving advice, persuasion, suggestion, and training in specific curative activities. Behaviour therapies are aimed at correcting specific pathological emotional states or behaviour patterns by appropriate countermeasures. They are based largely on the conditioned-reflex theory of I.P. Pavlov and on other theories of learning.

Individual therapies that aim to foster a patient's general personality growth emphasize helping him to gain insight into his feelings and behaviour. To facilitate this they try to create a therapeutic situation that will enable the patient to express himself with complete freedom, while the therapist maintains a consistent, warm, non-judgmental interest. Feeling himself understood and accepted by someone whom he admires and to whom he feels close, the patient will progressively dare to reveal those shameful or frightening aspects of himself that he has pushed out of awareness.

Some schools of psychotherapy hold that the consistent, warm "unconditional positive regard" of the therapist for the patient is sufficient to produce important changes. Therapies in the psychoanalytic tradition, while also emphasizing the importance of the therapeutic relationship, try to help the patient understand and master his feelings by analyzing them. They differ in their concepts and in the relative emphasis placed on different types of material produced by the patient.

Traditional psychoanalysis emphasizes the use of dreams as short cuts to the patient's deeper feelings. It also puts great stress on helping the patient to rediscover, re-experience, and "work through" the traumatic emotional experiences of early life in which his current difficulties are believed to originate. Later modifications of psychoanalysis put more emphasis on analysis of the patient's current problems, and some emphasize helping the patient to gain a better philosophy of life.

All agree that in an intimate, prolonged relation with the therapist, the patient will eventually experience toward him the feelings that trouble his relationships with persons emotionally close to him in his past and present life. Since both therapist and patient can observe these "transference reactions," as Freud termed them, exploring their inappropriateness is deemed a powerful means of resolving them.

There is no convincing evidence that the results of one form of treatment are better than any other. Despite differences in emphasis, most schools of psychotherapy agree that mental illnesses are, at least in part, expressions of chronic states of anxiety and frustration related to unresolved inner conflicts or unsuccessful ways of dealing with other persons. Though genetically or physiologically caused vulnerabilities might contribute to the difficulties of these patients, unfortunate early experiences with family members and other emotionally significant persons are believed to play a major role.

Chances of successful treatment are generally held to be related to the degree of the patient's emotional involvement in the treatment process. This is influenced by the intensity of his suffering and by his faith in the therapist and the treatment method. The patient's expectation of help is enhanced by the therapist's ability to convince the patient that he understands him intimately and is dedicated to his welfare. Personal qualities of the therapist seem important in the development of a successful therapeutic relationship.

NON-DIRECTIVE PSYCHOTHERAPY

This is also called *CLIENT-CENTRED PSYCHOTHERAPY*; an approach to the treatment of mental disorders that aims primarily to foster the patient's general personality growth by helping him gain insight into his feelings and behaviour. The function of the therapist is to extend consistent, warm "unconditional positive regard" toward the "client" (avoiding the negative connotations of "patient") and, by repeating and restating the client's own verbalized concerns, to enable the client to see himself more clearly and react more openly with the therapist and others. Pace, direction, and termination of therapy are controlled by the client; the therapist merely acts as a facilitator. The nondirective approach was originated by the American counselling psychologist Carl Rogers in the 1940s and influenced later individual and group psychotherapeutic methods.

DRUG THERAPY

The use of drugs to treat emotional disorders has expanded dramatically with the development of new and more effective medications for a variety of disorders that formerly were not treatable. Drugs that affect the mind are called psychotropic and can be divided into three categories: antipsychotic drugs, anti-anxiety agents, and antidepressant drugs.

ANTIPSYCHOTIC AGENTS

The advent of antipsychotic, or neuroleptic, drugs such as Thorazine (trademark) enabled many patients to leave mental hospitals and function in society. The primary indication for the use of anti-psychotics is schizophrenia, erroneously called split personality. This is a severe mental disorder characterized by delusions, hallucinations, and sometimes bizarre behaviour. One form, paranoid schizophrenia, is marked by delusions that are centred on a single theme, often accompanied by hallucinations. The most effective drug to use may depend on an individual patient's metabolism of the drug or the severity and nature of the side effects.

BRIEF FOCAL PSYCHOTHERAPY

This is a form of short-term dynamic therapy in which a time limit to the duration of the therapy is often agreed upon with the patient at the outset. Sessions lasting 30 to 60 minutes are held weekly for, typically, five to 15 weeks. At the beginning of treatment the therapist helps identify the patient's problem or problems, and these are made the focus of the treatment. The problem should be an important source of distress to the patient and it should be modifiable within the time limit.

The therapist is more active, directive, and confrontational than in long-term dynamic therapy and ensures that the patient keeps to the focus of treatment and is not diverted by subsidiary problems or concerns. Some therapists deliberately aim to produce considerable emotional arousal in the patient during each session as a way to activate or highlight specific problems. Research suggests that brief therapy can produce as good results as long-term therapy, and more quickly.

INDIVIDUAL DYNAMIC PSYCHOTHERAPY

Although psychoanalysis has had a profound influence, particularly on American psychiatry, that influence waned significantly during the 1970s and '80s. Fewer patients now enter psychoanalysis, and many analysts carry out short-term individual dynamic psychotherapy. This form of therapy is much more readily available and usually requires 50 minutes a week for six to 18 months. The aim of treatment, as in psychoanalysis, is to increase the patient's insight (understanding of himself), to relieve his symptoms, and to improve his psychological functioning. Suitable patients include those with a wide range of neurotic disorders and personal or social problems who wish to change and who are able to view their problems in psychological terms.

As in psychoanalysis, the patient learns to trust the therapist and becomes able to talk candidly and honestly about his most intimate thoughts and feelings. The treatment setting is less formal than in psychoanalysis, with the therapist and patient seated so that eye contact can be achieved if desired.

Treatment techniques include free association and the use of interpretation by the therapist to analyze the transference, the patient's unconscious defence mechanisms, and his dreams.

The therapist may ask the patient to clarify or enlarge on some point on which the therapist is not clear if this seems important in the development of the patient's symptoms.

The therapist directs the patient's attention to important links, of which he seems unaware, between the present and the past, between his emotional responses to the therapist and to people important to him, and so on.

The therapist may challenge the patient with the likely consequences of his resistant or maladaptive behaviour and stress instead the importance of confronting and trying to resolve his psychological difficulties.

4. DEVELOPMENT OF PSYCHOTHERAPY

Psychotherapy implies the treatment of mental discomfort, dysfunction, or disease by psychological means by a trained therapist who adheres to a particular theory of both symptom causation and symptom relief. However, the many forms of psychotherapy have been developed by modern medicine and which are carried out by a member of one of the mental health professions such as a psychiatrist or a clinical psychologist.

Foremost among these was that of psychoanalysis, which originated in the work of the Viennese neurologist Sigmund Freud. Having studied under the French neurologist Jean-Martin Charcot, Freud originally used well-known techniques of hypnosis to treat patients suffering from hysterical paralysis and other neurotic syndromes. Freud and his colleague, Josef Breuer, observed that their patients tended to relive earlier life experiences that could be associated with the symptomatic expression of their illnesses.

When these memories and the emotions associated with them were brought to consciousness during the hypnotic state, the patients showed improvement. Observing that most of his patients proved able to talk about such memories without being under hypnosis, Freud evolved the technique of free association (the production by the patient, aloud and without suppression or self-censorship of any kind, of the thoughts and feelings about whatever was uppermost in his mind) as a means of access to the unconscious.

From this beginning Freud gradually developed what became known as psychoanalysis. Other features of the new procedure included the study of dreams, the interpretation of "resistances" on the part of the patient, and the handling by the therapist of transference (the patient's feelings toward the analyst that reflected previously experienced feelings toward parents and other important figures in his early life).

Freud's work, though complex and controversial in many of its aspects, laid the basis for modern psychotherapy in its use of free association and its emphasis on unconscious and irrational mental processes as causative factors in mental illness. This emphasis on purely psychological factors as a basis for both causation and treatment was to become the cornerstone of most subsequent psychotherapies.

Variations of the original psychoanalytic technique were introduced by several of Freud's colleagues who parted company with him. Analytic psychology, devised by Carl Jung , placed less emphasis on free association and more on the interpretation of dreams and fantasies.

Special importance was given to the collective unconscious, a reservoir of shared unconscious wisdom and ancestral experience that entered consciousness only in symbolic form to influence thought and behaviour.

Jungian analysts sought clues to their patients' problems in the archetypal nature of myths, stories, and dreams. Individual psychology, devised by Alfred Adler, emphasized the importance of the individual's drive toward power and of his unconscious feelings of inferiority. The therapist was concerned with the patient's compensations for his inferiority, as well as with his social relationships.

MENTAL DISORDER TREATMENT

The most successful treatment approaches combine the use of drugs, psychotherapy, and supportive therapy. In acute schizophrenia, phenothiazine, chlorpromazine, or butyrophenone drugs such as haloperidol are of proven efficacy in relieving or eliminating such symptoms as delusions, hallucinations, thought disorders, agitation, and violent behaviour. Long-term maintenance on such drugs also reduces the rate of relapse.

Psychotherapy serves to relieve the patient's feelings of helplessness and isolation, buttress his healthy or positive tendencies, and help him to distinguish between his psychotic perceptions and reality and to deal with any underlying emotional conflicts that might be exacerbating his condition. Occupational therapy for those in day care and regular visits from a social worker or community psychiatric nurse for outpatients are beneficial. It is sometimes useful to counsel the relatives of schizophrenic patients living at home in their way of dealing with the patient's symptoms.

PSYCHODYNAMIC THERAPIES

Sigmund Freud held that all behaviour is influenced by unconscious motivations and conflicts. Personality characteristics are thought to be shaped from the earliest childhood experiences. Psychological defences are seen mainly as unconscious coping responses, the purpose of which is to resolve the conflicts that arise between basic desires and the constraints of external reality. Emotional problems are seen as maladaptive responses to these unconscious conflicts.

Psychodynamic therapies emphasize that insight is essential to lasting change. Insight means understanding how a problem emerged and what defensive purpose it serves. A classic form of psychodynamic therapy is psychoanalysis, in which the patient engages in free association of ideas and feelings and the psychoanalyst offers interpretations as to the meaning of the associations.

Another form is brief, dynamic psychotherapy, in which the clinician makes recommendations based on an understanding of the situation and the reasons for resisting change.

Psychotherapy is used for mental rather than physical means to achieve behavioural or attitudinal change, employs suggestion, persuasion, education, reassurance, insight, and hypnosis. Supportive psychotherapy is used to reinforce a patient's defences, but avoids the intensive probing of emotional conflicts employed in psychoanalysis and intensive psychotherapy.

Experienced clinicians usually draw on various counselling theories and techniques to design interventions that fit a patient's problem. The format of therapy (*e.g.*, individual, couple, family, or group) will vary with each patient. Many patients respond best to a combination approach.

Depression, for example, is frequently alleviated by medication and cognitive-behavioural therapy. There is growing interest in primary prevention to increase the coping abilities and resilience of children, families, and adults who are at risk for mental health problems.

AVERSION THERAPY

Aversion therapy in psychotherapy is designed to cause a patient to reduce or avoid an undesirable behaviour pattern by conditioning him to associate the behaviour with an undesirable stimulus. The chief stimuli used in the therapy are electrical and chemical.

In the electrical therapy, the patient is given a lightly painful shock whenever the undesirable behaviour is aroused; this method has been used in the treatment of sexual deviations.

In the chemical therapy, the patient is given a drug that produces unpleasant effects, such as nausea, when combined with the undesirable behaviour; this method has been common in the treatment of alcoholism, the therapeutic drug and the alcohol together causing the nausea.

PSYCHOANALYTIC PSYCHOTHERAPY

Classical psychoanalysis is the most demanding of all the psychotherapies in terms of time, cost, and effort. It is conducted with the patient lying on a couch and with the analyst seated out of his sight but close enough to hear what the patient says.

The treatment sessions last 50 minutes and are usually held four or five times a week for at least three years. The primary technique used in psychoanalysis and in other dynamic psychotherapies to enable unconscious material to enter the patient's consciousness is that of "free association."

In free association, according to Freud, the patient "is to tell us not only what he can say intentionally and willingly, what will give him relief like a confession, but everything else as well that his self-observation yields him, everything that comes into his head, even if it is disagreeable for him to say it, even if it seems to him unimportant or actually nonsensical."

Such a procedure is rendered difficult, first, because for a person to speak of his innermost (and often socially unacceptable) thoughts is a departure from years of practice in which he has selected what he has said to others.

Free association is also difficult because the patient resists remembering repressed experiences or feelings that are connected with intense or conflicting emotions that have never been finally resolved or settled.

Such repressed emotions or memories usually revolve around the patient's important personal relationships and his innermost feelings of self, and the release or recollection of such emotions in the course of treatment can be itself intensely disturbing.

Attentive listening and "empathy" on the part of the therapist allows the patient to express thoughts and feelings that in turn permit the uncovering of his underlying emotional conflicts.

In the course of treatment, the patient often seeks to project (attribute to something other than himself) the disturbing emotions he feels in the process of recollection and free association, and the person who is almost invariably selected for the focus of such projection is the psychoanalyst; that is, the patient is likely to blame his emotional distress on the analyst. In this way, the patient comes to feel love or hatred, dependence or rebellion, and rivalry or rejection toward the analyst.

These are the same attitudes the patient has felt but has never consciously acknowledged toward his parents or other people with whom he shared important relations earlier in life.

The patient's projection onto the therapist of these feelings and behaviours that originated in his earlier relationships is called the transference. To facilitate the development of the transference, the analyst endeavours to maintain a neutral stance toward the patient in order to serve as a "blank screen" onto which the patient can project his inner feelings.

The analyst's handling of the transference situation is of vital importance in psychoanalysis or, indeed, in any form of dynamic psychotherapy. It is through the transference that the patient discovers the nature of his unconscious feelings and then becomes able to acknowledge them. Once this has been done, he often finds himself able to regard them in a far more dispassionate and tolerant light and often feels himself liberated from their influence upon his future behaviour.

INTERPRETATION

A major therapeutic tool in the course of treatment is interpretation. This technique helps the patient to become aware of previously repressed aspects of his emotional conflicts and to uncover the meaning of uncomfortable feelings evoked by the transference.

Interpretation, in turn, is used to determine the underlying psychological meaning of the patient's dreams, which are held to have a hidden or latent content that symbolize and indirectly express aspects of the patient's emotional conflicts.

5. TWENTY-FIRST CENTURY MENTAL HEALTH

There has been a great increase in the number of mental health professionals since World War II. In the United States the number of psychiatrists was 3,000 in 1939 but had increased to more than 25,000 by the early 1970s. Non-medical mental health professionals have also increased in number and have achieved increasing independence from medical control, acquiring new roles in the process.

Clinical psychologists, for instance, who were at one time largely confined to carrying out psychometric tests at the doctor's request, have become increasingly concerned with psychotherapy and behaviour therapy.

Psychiatric social workers are no longer confined to casework with individual patients or their relatives but have also become psychotherapists and play prominent roles in mental health centres.

There are new roles for nurses, including behaviour therapy and the management of chronic mental illness in the community. The greatest beneficiaries of this expansion of mental health professionals have been patients with neurotic and other less severe disorders.

Psychotherapy retains a major role in the mental health profession, and since the development of psychoanalysis the varieties of psychotherapy have increased and multiplied to the extent that a 1980 handbook listed "more than 250 different therapies in use today."

The repertoire of drugs used in the treatment of mental illness has continued to grow as new drugs are developed or new applications of existing ones are discovered, and research on the biochemical and genetic causes of mental disease continues to make gradual headway in explicating the causes of various disorders.

The triad of psychotherapy, drugs, and behaviour therapy afford an unprecedented array of approaches, techniques, and procedures for alleviating symptoms or curing people altogether of mental disorders.

GROUP PSYCHOTHERAPY

Many types of psychological treatment may be provided for groups of patients with psychiatric disorders. This is true, for example, of relaxation training and anxiety-management training.

There are also self-help groups, of which Alcoholics Anonymous is perhaps the best known. A considerable number of group experiences have been devised for people who are not suffering from any psychiatric disorder; encounter groups are a well-known example.

This discussion, however, is concerned with long-term dynamic group therapy, in which six to 10 psychiatric patients meet with a trained group therapist, or sometimes two therapists, for 60 to 90 minutes a week for up to 18 months.

Often the group is closed, *i.e.,* confined to the original group membership, even if one or more members drop out before the treatment ends. In an open group patients who have stopped attending, whether by default or because of the relief of symptoms, are replaced by new members.

The types of mental disorders considered suitable for group therapy are much the same as those suitable for individual therapy. Again the patient must want to change and must be psychologically minded. In addition, it is important that he not consider group therapy as a poor second to individual therapy.

There are many varieties of dynamic group therapy, and they differ in their theoretical background and technique. The influential model of the American psychiatrist Irvin D. Yalom provides a good example of such therapies, however.

The therapist continually encourages the patients to direct their attention to the personal interactions occurring within the group rather than to what happened in the past to individual members or what is currently happening outside the group, although both of these areas may be considered when they are relevant.

The therapist regularly draws attention to what is happening among members of the group as they learn more about themselves and test out different ways of behaving with one another. The goal in group therapy is to create a climate in which the participants can shed their inhibitions.

When the members come to trust one another, they are able to provide feedback and to respond to other group members in ways that might not be possible in ordinary social interactions owing to the constraints of social conventions.

Several factors appear to be important in group therapy. The most important is group cohesion, which gives the patient a feeling of belonging, identification, and security and enables him to be frank and open and to take risks without the danger of rejection.

Universality refers to the patient's realization that he is not unique, that all the other group members have problems, some of them similar to his. Optimism about what can be achieved in the group, fostered by the perception of change in others, combats demoralization.

Guidance, the giving of advice and explanation, is important in the early meetings of the group and is largely a function of the therapist. What has been called vicarious learning later becomes more important; the patient observes how other group members evolve solutions to common problems and emulates desirable qualities he sees in fellow members.

Catharsis, or the release of highly charged emotion, occurs within the group setting and is helpful provided that the patient can come to understand it and appreciate its significance. Another factor that is helpful in improving self-esteem is altruism, the opportunity to give assistance to another group member.

6. ESTABLISHED THERAPIES

Many other types of psychotherapy have been developed in the second half of the 20th century, each with its own emphasis on symptom causation and its own particular approach to treatment.

Many of these therapies use classical dynamic and behavioural models in modified forms, and they may also stress the understanding and modification of cognition and the ways in which people "process" their experiences, moods, and emotions.

Among these relatively recent psychotherapies are client-centred psychotherapy, developed by the American psychologist Carl R. Rogers; transactional analysis, originated by the American psychiatrist Eric Berne; the interpersonal therapies developed by the American psychiatrists Adolf Meyer and Harry Stack Sullivan; cognitive therapy, developed by the American psychiatrist Aaron T. Beck; rational-emotive psychotherapy, developed by the American psychologist Albert Ellis; and Gestalt therapy, which stems from the work of the German psychiatrist Frederick S. (Fritz) Perls.

Another class of therapies consists of those used to care for psychotic patients, both those in hospitals and those who live in the community.

Supportive psychotherapy consists of the long-term help of patients who are chronically handicapped by schizophrenia or other mental disorders. Such a therapy uses reassurance, guidance, and encouragement to help the patient cope with his disabilities and live as satisfactory a life as possible.

Rehabilitation programs for chronic or episodically psychotic patients include drug maintenance; training in social skills that they may have lost while sick; occupational training to improve the patient's skills in cooking, shopping, and other domestic tasks; and industrial therapy, which usually offers the patient gainful employment under conditions of minimal stress.

Family therapy is sometimes used to help relatives learn to cope with a schizophrenic patient who has returned home from the hospital.

Community care for released schizophrenics or other psychotic patients must provide them with drug maintenance and a minimum of psychiatric monitoring; appropriate housing facilities; some type of employment; and training in such skills as using public transport, preparing their own food, and looking after their finances.

Each patient should have a case manager, a professional worker who maintains contact and secures from governmental or social agencies the assistance that the patient needs.

When provisions like these are not made, some formerly hospitalized patients stop taking their medicine and in effect drop out of the mental health care system, becoming unemployed and even homeless.

This phenomenon became particularly evident in the United States and to a lesser extent in Western Europe, when massive numbers of mental patients were released from hospitals during the 1950s and '60s after the effectiveness of antipsychotic drugs had been verified.

These releases were also motivated by the concerns of civil libertarians over the abuse of patients' rights in keeping them committed to mental hospitals.

However, the support network of community-based mental health clinics that would have been necessary to cope with the released patients was either inadequately established or nonexistent.

The result was that many psychotic patients received inadequate outpatient care and supervision or encountered severe difficulties in obtaining housing or employment, becoming homeless wanderers in large urban areas.

DYNAMIC PSYCHOTHERAPIES

There are many variants of dynamic psychotherapy, all of which ultimately derive from the basic precepts of psychoanalysis.

The fundamental approach of most dynamic psychotherapies can be traced to three basic theoretical principles or assertions:

(1) Human behaviour is prompted chiefly by emotional considerations, but insight and self-understanding are necessary to modify and control such behaviour and its underlying aims;

(2) A significant proportion of human emotion is not normally accessible to one's personal awareness or introspection, being rooted in the unconscious, those portions of the mind beneath the level of consciousness;

(3) Any process that makes available to a person's conscious awareness the true significance of emotional conflicts and

tensions that were hitherto held in the unconscious will thereby produce heightened awareness and increased stability and emotional control.

The classic dynamic psychotherapies are relatively intensive talking treatments that are aimed at providing the person with insight into his conscious and unconscious mental processes, with the goal of enabling him ultimately to achieve a better understanding of himself.

Dynamic psychotherapy attempts to enhance the patient's personality growth as well as to alleviate his symptoms. The main therapeutic forces are activated in the relationship between patient and therapist and depend both upon the empathy, understanding, integrity, and concern demonstrated by the therapist and upon the motivation, intelligence, and capacity for achieving insight manifested by the patient.

The attainment of a therapeutic alliance--*i.e.*, a working relationship between patient and therapist that is based on mutual respect, trust, and confidence--provides the context in which the patient's problems can be worked through and resolved. Several of the most important forms are treated below.

SENSITIVITY TRAINING

This is a psychological technique in which intensive group discussion and interaction are used to increase individual awareness of self and others; it is practiced in a variety of forms under such names as T-group, encounter group, human relations, and group-dynamics training. The group is usually small and unstructured and chooses its own goals.

A trained leader is generally present to help maintain a psychologically safe atmosphere in which participants feel free to express themselves and experiment with new ways of dealing with others.

The leader remains as much as possible outside the discussion. Issues are raised by the group members, and their interactions evoke a wide variety of feelings. The leader encourages participants to examine verbally their own and others' reactions.

It is believed that as mutual trust is developed, interpersonal communication increases, and eventually attitudes will change and be carried over into relations outside the group.

Often, however, these changes do not endure. Sensitivity training seems to be most effective if sessions are concentrated and uninterrupted, as in several days of continuous meetings.

Sensitivity-training methods derived in large part from those of group psychotherapy. They have been applied to a wide range of social problems (as in business and industry) in an effort to enhance trust and communication among individuals and groups throughout an organization.

SOCIAL PSYCHOLOGY AND INTERACTION PROCESSES

The different verbal and non-verbal signals used in conversation have been studied, and the functions of such factors as gaze, gesture, and tone of voice are analyzed in social-interaction studies.

Social interaction is thus seen to consist of closely related sequences of nonverbal signals and verbal utterances. Gaze has been found to perform several important functions.

Laboratory and field studies have examined helping behaviour, imitation, friendship formation, and social interaction in psychotherapy.

Among the theoretical models developed to describe the nature of social behaviour, the stimulus-response model (in which every social act is seen as a response to the preceding act of another individual) has been generally found helpful but incomplete.

Linguistic models that view social behaviour as being governed by principles analogous to the rules of a game or specifically to the grammar of a language have also attracted adherents.

Others see social behaviour as a kind of motor skill that is goal-directed and modified by feedback (or learning), while other models have been based on the theory of games, which emphasizes the pursuit and exchange of rewards and has led to experiments based on laboratory games.

GESTALT THERAPY

A humanistic method of psychotherapy that takes a holistic approach to human experience by stressing individual responsibility and awareness of present psychological and physical needs.

Frederick (Fritz) S. Perls, a German-born psychiatrist, founded Gestalt therapy in the 1940s with his wife, Laura. Perls was trained in traditional psychoanalysis, but his dissatisfaction with certain Freudian theories and methods led him to develop his own system of psychotherapy. He was influenced by the psychoanalysts Karen Horney and Wilhelm Reich.

Also influential were ideas expressed in existentialism and phenomenology, such as freedom and responsibility, the immediacy of experience, and an individual's role in creating meaning in his life.

Gestalt psychology provided a framework for Perls' system. According to this psychology of perception, when organisms are confronted with a set of elements, they perceive a whole pattern or configuration, rather than bits and pieces, against a background.

Perls applied this concept to human experience, postulating that healthy persons organize their field of experience into well-defined needs to which they respond appropriately.

For example, when various perceptions lead a healthy person to experience the Gestalt of hunger, he eats. On the other hand, a neurotic interferes with the formation of the appropriate Gestalt and does not adequately deal with his need.

In another example of an unhealthy response, a person who has just received an insult may be angry but may partially or completely repress his awareness of his anger.

Gestalt therapy seeks to resolve the conflicts and ambiguities that result from the failure to integrate features of the personality. The goal of Gestalt therapy is to teach people to become aware of significant sensations within themselves and their environment so that they respond fully and reasonably to situations.

The focus is on the "here and now," rather than on past experiences, although once the client becomes aware of the present he can confront past conflicts or unfinished business--what Perls referred to as incomplete Gestalts.

Clients are urged to discuss their memories and concerns in the present tense. Dramatizing conflicts is also a method Gestalt therapists use to make problems understandable to their clients; they may be called on to act out repressed aspects of their personalities or to adopt the role of other individuals.

Like other humanistic therapies, Gestalt therapy assumes the innate inclination of individuals to health, wholeness, and realization of their potential. The awareness it seeks to cultivate is essential to restoring the natural inclination to happiness and fulfilment.

Perls developed most of the techniques of Gestalt therapy in the United States. He helped establish Gestalt institutes in many parts of the country, including Esalen in Big Sur, California, U.S., during the 1960s.

Many of his techniques have been incorporated into eclectic approaches to psychotherapy.

HOLISTIC MEDICINE

A doctrine of preventive and therapeutic medicine that emphasizes the necessity of looking at the whole person--his body, mind, emotions, and environment--rather than at an isolated function or organ and which promotes the use of a wide range of health practices and therapies.

It has especially come to stress responsibility for "self-healing," or "self-care," by observing the traditional commonsense essentials of exercise, healthful diet, adequate sleep, good air, moderation in personal habits, and so forth.

The term holistic medicine became especially fashionable in the late 20th century (the International Association of Holistic Health Practitioners was founded in 1970, assuming its current holistic name in 1981).

In its underlying philosophy, in emphasizing the provision of whole care to a person or patient, holistic medicine is not new, being inseparable from any traditional health care of good quality.

Holistic medicine in extreme instances, however, has tended to equate the validity of a wide range of schools or approaches to health care, not all of them compatible and some of them competitive, some scientific and some unscientific.

Although mainstream Western medical practices are not ignored, they are seen as only one part of the available therapies and by no means the only effective ones.

Congresses and conferences on holistic health have thus drawn not only representatives of medical schools and institutions but also advocates of such widely varying concepts as acupuncture, alternative childbirth, astrology, biofeedback, chiropractic, faith healing, graphology, homeopathy, macrobiotics, megavitamin therapy, naturopathy, numerology, nutrition, osteopathy, psychocalisthenics, psychotherapy, self-massage, shiatsu (or acupressure), touch encounter, and yoga.

ANDREAS SOFRONIOU

BRAINWASHING AND COERCIVE PERSUASION

This is a systematic effort to persuade non-believers to accept a certain allegiance, command, or doctrine. A colloquial term, it is more generally applied to any technique designed to manipulate human thought or action against the desire, will, or knowledge of the individual.

By controlling the physical and social environment, an attempt is made to destroy loyalties to any unfavourable groups or individuals, to demonstrate to the individual that his attitudes and patterns of thinking are incorrect and must be changed, and to develop loyalty and unquestioning obedience to the ruling party.

The term is most appropriately used in reference to a program of political or religious indoctrination or ideological remoulding. The techniques of brainwashing typically involve isolation from former associates and sources of information; an exacting regimen requiring absolute obedience and humility; strong social pressures and rewards for cooperation; physical and psychological punishments for non-cooperation ranging from social ostracism and criticism, deprivation of food, sleep, and social contacts, to bondage and torture; and continual reinforcement.

The nature of brainwashing as it occurred in communist political prisons received widespread attention after the Chinese Communist victory in 1949 and after the Korean and Vietnamese wars. More recently, its reported use in fringe religious cults and radical political groups has aroused concern in the United States.

Deprogramming, or reversing the effects of brainwashing through intensive psychotherapy and confrontation, has proved somewhat successful, particularly with religious cult members.

SEX THERAPY

Sex therapy is a form of behaviour modification or psychotherapy directed specifically at difficulties in sexual interaction. Many sex therapists use techniques developed in the 1960s by the Americans William Masters and Virginia Johnson to help couples with non-organic problems that affect their sex lives, including premature ejaculation, impotence, and other forms of sexual dysfunction. In the Masters and Johnson technique, a sex history is first taken and the couple given physical examinations to rule out physical problems.

Therapists then employ exercises focusing on the giving and receiving of sensual, but not necessarily sexual, pleasure to help the couple overcome anxieties about sex. Specialized treatments directed against specific sex-related problems are also used during therapy. The therapy process often takes place in an intensive marital workshop lasting for several days.

Although the Masters and Johnson approach involves both members of the couple, sex therapy can take other forms. Co-marital therapy refers to the Masters and Johnson model, in which both members of the couple are treated by a team consisting of one male and one female therapist.

The couple approach recognizes that sexual dysfunctions take place in the context of the interaction between two people and are not the exclusive problem of one member of the pair.

Individual therapy is employed for those without cooperating partners and may involve the use of a surrogate partner or may focus on exercises that can be practiced by an individual to improve his or her sexual interactions. Group therapy, in which individuals discuss feelings about sex, is also employed for both single-sex and male-female groups.

TREATMENT WITH MUSIC

Music is the art concerned with combining vocal or instrumental sounds for beauty of form or emotional expression, usually according to cultural standards of rhythm, melody, and, in most Western music, harmony.

Both the simple folk song and the complex electronic composition belong to the same activity, music. Both are humanly engineered; both are conceptual and auditory, and these factors have been present in music of all styles and in all periods of history, Eastern and Western.

Music is an art that, in one guise or another, permeates every human society. Modern music is heard in a bewildering profusion of styles, many of them contemporary, others engendered in past eras. Music is a protean art; it lends itself easily to alliances with words, as in song, and with physical movement, as in dance.

Throughout history, music has been an important adjunct to ritual and drama and has been credited with the capacity to reflect and influence human emotion.

Popular culture has consistently exploited these possibilities, most conspicuously today by means of radio, film, television, and the musical theatre. The implications of the uses of music in psychotherapy, geriatrics, and advertising testify to a faith in its power to affect human behaviour.

Publications and recordings have effectively internationalized music in its most significant, as well as its most trivial, manifestations. Beyond all this, the teaching of music in primary and secondary schools has now attained virtually worldwide acceptance.

But the prevalence of music is nothing new, and its human importance has often been acknowledged. What seems curious is that, despite the universality of the art, no one until recent times has argued for its necessity.

The ancient Greek philosopher Democritus explicitly denied any fundamental need for music: "For it was not necessity that separated it off, but it arose from the existing superfluity."

The view that music and the other arts are mere graces is still widespread, although the growth of psychological understanding of play and other symbolic activities has begun to weaken this tenacious belief.

7. PHARMACOLOGICAL TREATMENTS

Specific treatment for mental disorders involves antipsychotic agents and other pharmacological treatments (also known as neuroleptics). These with a few other major tranquilizers, belong to several different chemical groups but are similar in their therapeutic effects. They have a calming effect that is valuable in the relief of agitation, excitement, and violent behaviour in psychotic patients.

The drugs are quite successful in reducing the symptoms of schizophrenia, mania, and delirium, and they are used in combination with antidepressants to treat psychotic depression. The drugs suppress hallucinations and delusions, alleviate disordered or disorganized thinking, improve the patient's lucidity, and generally make him more receptive to psychotherapy.

Patients who have previously been agitated, intractable, or grossly delusional become noticeably calmer, quieter, and more rational when maintained on these drugs. The drugs have enabled many episodically psychotic patients to have shorter stays in hospitals, and many other patients who would have been permanently confined to institutions are able to live in the outside world because of the drugs.

The anti-psychotics differ in their unwanted effects: some are more likely to make the patient drowsy; some to alter blood pressure or heart rate; and some to cause tremor or slowness of movement.

In the treatment of schizophrenia, antipsychotic drugs partially or completely control such symptoms as delusions and hallucinations. They also protect the patient who has recovered from an acute episode from suffering a relapse.

Unfortunately, the drugs have less of an effect on such symptoms as social withdrawal, apathy, blunted emotional capacity, and the other psychological deficits characteristic of the chronic stage of the illness.

No single drug seems to be outstanding in the treatment of schizophrenia. In an individual patient, one drug may be preferred to another because it produces less severe unwanted effects, and the dose of any one drug needed to produce a therapeutic effect varies widely from patient to patient.

Because of these individual differences it is common for psychiatrists to substitute a drug of a different chemical group when one drug has been shown to be ineffective despite its use in adequate dosage for several weeks.

In an acute psychotic episode a drug such as chlorpromazine, trifluoperazine, or haloperidol usually has a calming effect within a day or two. The control of psychotic symptoms such as hallucinations or disordered thinking may take weeks. The appropriate dosage has to be determined for each patient by cautiously increasing the dose until a therapeutic effect is achieved without unacceptable side effects.

Antipsychotic drugs are thought to work by blocking dopamine receptors in the brain. Dopamine is a neurotransmitter; *i.e.*, a chemical messenger produced by certain nerve cells that influence the function of other nerve cells by interacting with receptors in their cell membranes.

Since schizophrenia may be caused by either the excessive release of or an increased sensitivity to dopamine in the brain, the effects of antipsychotic drugs may be due to their ability to block or inhibit dopamine transmission.

Dopamine receptor blockade is responsible for the drug's main unwanted side effects, which are termed extra-pyramidal symptoms (EPS). These resemble the symptoms of Parkinson's disease and include tremor (shakiness) of the limbs; bradykinesia--slowness of movement with loss of facial expression, absence of arm-swinging during walking, and a general muscular rigidity; dystonia--sudden, sustained contraction of muscle groups causing abnormal postures; akathisia--a subjective feeling of restlessness leading to an inability to keep still; and tardive dyskinesia--involuntary movements, particularly involving the lips and tongue.

Most extrapyramidal symptoms disappear when the drug is withdrawn. Tardive dyskinesia occurs late in the drug treatment and in about half of the cases persists even after the drug is no longer used. There is no satisfactory treatment.

8. DESCRIPTION OF PSYCHOLOGICAL SYMPTOMS

KLEPTOMANIA

This refers to recurrent compulsion to steal without regard to the value or use of the objects stolen. Although widely known and sometimes used as an attempted legal defence by arrested thieves, genuine kleptomania is a fairly rare mental disorder. A kleptomaniac may hide, give away, or secretly return the stolen items, but he seldom uses them or attempts to profit by their resale. The kleptomaniac usually has the economic means to purchase what he steals and obtains gratification from the theft itself rather than from its object.

Kleptomania is classified as a disorder of impulse control, meaning that the victim is unable to overcome the urge to steal and feels an increasing tension with attempts to resist, until yielding to the impulse gives release. In some cases, the stolen objects may have a symbolic sexual or other significance for the kleptomaniac, but the sexual aspects of the disorder are not always evident. Psychotherapy can be effective in alleviating the disorder, but few kleptomaniacs seek help unless they are caught in a theft or are referred to a psychiatrist for treatment of depression or anxiety related to their fears of being apprehended.

BULIMIA

This mental disorder is characterized by periods of binging (extreme overindulgence in food) followed by purging, including induced vomiting. Repetition of the cycle can lead to serious medical complications, such as dental decay, stomach rupture, or dehydration, and can be fatal.

Bulimia generally begins during adolescence or early adult life and is more prevalent in females. Individuals who suffer from this disorder show great concern for their weight and body shape; the majority, however, are close to their proper weight. Depression, anxiety, and low self-esteem are associated features. Treatment may include psychotherapy. The disorder is distinct from anorexia nervosa, which involves an emotional or psychological aversion to food and results in extreme emaciation.

FANTASY

Fantasy is a type of a daydream arising from conscious or unconscious wishes or attitudes. Theorizing about fantasy and imagination goes back at least to Aristotle, but proper psychological interest in it has been a recent phenomenon, partly due to behaviourism's denial of the possibility of studying mental content and partly perhaps because it was not taken seriously. Recent work suggests that fantasy and visualization are valuable human capacities.

Children who engage in fantasy concentrate better, have more positive emotions, and often cope better with relationships. Fantasy visualization has been used in various psychotherapies. Patients fantasize to induce mild anxiety with which relaxation techniques help them to deal, or they use attractive fantasies as part of a relaxation procedure. Sports psychologists encourage their charges to visualize themselves succeeding so as to build up their confidence. Visualizing one's plans as they might appear to others may give a more realistic estimate of what one might do.

ANXIETY

Anxiety is a feeling of actual or anticipated unease, similar to fear, but more closely associated with uncertainty. It is associated with bodily changes such as raised heart-rate and perspiration. The relationship between anxiety and uncertainty is shown by the fact that people who are unsure whether they will receive an electric shock have a higher heart-rate and perspire more than those who are sure that a shock is coming. Psychologists distinguish the state of being anxious from the trait of being an anxious person. Levels of trait anxiety appear to be major, and perhaps innate, differences between people.

In psychiatry, anxiety occurs in many psychological disorders, both neuroses and psychoses. A notable exception is the psychopathic personality, where it is precisely the absence of anxiety (failure to worry about the consequences of one's actions) that is the problem. In certain disorders, such as phobias, in which extreme anxiety is the main feature, the person is incapacitated by fear of, and hence driven to avoid, what in reality are harmless situations.

Although drugs (minor tranquillizers) can be used in the treatment of abnormal anxiety, this may lead to dependence and does not relieve the underlying problem, which is better dealt with by psychotherapy or one of the newer instructional methods (behaviour therapy), whereby the person is taught self-help techniques of anxiety management. A related term is angst, first used by Kierkegaard to describe an existential anxiety or anguish.

DISABILITY

Although a physical ailment, disability can also be a mental incapacity. The World Health Organization (WHO) has estimated that 10 per cent of the world's population is disabled in some way, despite the fact that half of all disabilities are preventable. Causes include traffic accidents, war injuries, infectious diseases, malnutrition, psychiatric illness, and degenerative diseases. WHO has developed a conceptual framework for the assessment of individual and population needs.

Impairment is defined as a loss or abnormality of an anatomical, physiological, or psychological function, such as the loss of a leg, vision, or mental functioning. Disability is defined as a partial or complete inability, arising from impairment, to perform an activity in the manner, or within the range, considered normal for a human being. Examples would be inability to walk or to care for oneself.

Handicap is defined as a disadvantage, resulting from impairment or disability, which limits or prevents fulfilment of a role that is normal, depending on age and social factors.

Thus impairments and disabilities need not necessarily turn into handicaps. This depends on the individual concerned and, critically, on social responses: inability to walk would not prevent a lawyer from working, if his or her colleagues and clients did not discriminate, and if the office were accessible for a wheelchair.

Disability prevention can be related to the WHO classification by reducing the occurrence of impairments, limiting or reversing disability, and preventing a disability from becoming a handicap.

IMPOTENCE

Impotence is inability of the male to have satisfactory sexual intercourse and varies in form from the inability to gain an erection to weak erections, premature ejaculation, or loss of normal sensation with ejaculation.

Almost all of these complaints are psychogenic in origin, but impotence may be caused by subnormal functioning of the testes, by arteriosclerosis (hardening of the arteries), diabetes mellitus (a metabolic disease in which there is inadequate secretion or utilization of insulin), or by some disease of the nervous system.

Certain medications prescribed for the treatment of such diseases as peptic ulcer, hypertension, or psychiatric illnesses may adversely affect sexual ability.

Therapy, usually limited in its success, includes administration of sex hormones and psychotherapy.

NARCISSISM

In the narcissistic personality disorder, there is a grandiose sense of self-importance and a preoccupation with fantasies of success, power, and achievement.

Avoidant personalities are excessively sensitive to social rejection, humiliation, and shame, have low self-esteem, and are deeply upset by the slightest disapproval of others; they are consequently unwilling to enter into relationships but crave affection and acceptance.

Passive-aggressive personality disorder is the term applied to people who respond aggressively and negatively to demands made upon them by using such passive means as procrastination, dawdling, intentional inefficiency, or deliberate forgetfulness.

PERSONALITY TRAITS

Personality traits are, by definition, virtually permanent, and so these disorders are only partially, if at all, amenable to treatment.

The most effective treatment combines various types of group, behavioural, and cognitive psychotherapy.

The behavioural manifestations of personality disorders often tend to diminish in their intensity in middle and old age.

MELANCHOLIA/DEPRESSION

Depression in psychology is a mood or emotional state that is marked by sadness, inactivity, and a reduced ability to enjoy life.

A person who is depressed usually experiences one or more of the following symptoms: feelings of sadness, hopelessness, or pessimism; lowered self-esteem and heightened self-depreciation; a decrease or loss of ability to enjoy daily life; reduced energy and vitality; slowness of thought or action; loss of appetite; and disturbed sleep or insomnia.

Depression differs from simple grief, bereavement, or mourning, which are appropriate emotional responses to the loss of loved persons or objects. Where there are clear grounds for a person's unhappiness, depression is considered to be present if the depressed mood is disproportionately long or severe vis-à-vis the precipitating event.

When a person experiences alternating states of depression and mania (extreme elation of mood), he is said to suffer from a manic-depressive psychosis.

Depression is probably the most common psychiatric complaint and has been described by physicians from at least the time of Hippocrates, who called it melancholia. The course of the disorder is extremely variable from person to person; it may be fleeting or permanent, mild or severe, acute or chronic.

Depression is more common in women than in men. The rates of incidence of the disorder increase with age in men, while the peak for women is between the ages of 35 and 45.

Depression can have many causes. The loss of one's parents or other childhood traumas and privations can increase a person's vulnerability to depression later in life. Stressful life events in general are potent precipitating causes of the illness, but it seems that both psychosocial and biochemical mechanisms can be important causes.

The chief biochemical cause seems to be the defective regulation of the release of one or more naturally occurring monoamines in the brain, particularly norepinephrine and serotonin. Reduced quantities or reduced activity of these chemicals in the brain is thought to cause the depressed mood in some sufferers.

There are three main treatments for depression. The two most important are psychotherapy and drug therapy. Psychotherapy aims to resolve any underlying psychic conflicts that may be causing the depressed state, while also giving emotional support to the patient.

Antidepressant drugs, by contrast, directly affect the chemistry of the brain, and presumably achieve their therapeutic effects by correcting the chemical imbalance that is causing the depression.

The tricyclic antidepressant drugs are thought to work by inhibiting the body's physiological inactivation of the monoamine neurotransmitters. This results in the build-up or accumulation of these neurotransmitters in the brain and allows them to remain in contact with nerve cell receptors there longer, thus helping to elevate the patient's mood.

By contrast, the antidepressant drugs known as monoamine oxidase inhibitors interfere with the activity of monoamine oxidase, an enzyme that is known to be involved in the breakdown of norepinephrine and serotonin.

In cases of severe depression in which therapeutic results are needed quickly, electroconvulsive therapy has proven helpful. In this procedure, a convulsion is produced by passing an electric current through the person's brain.

INSOMNIA

This is the inability to sleep adequately. Insomnia varies greatly in pattern and degree; the responsible causes may include poor sleeping conditions, circulatory or brain disorders, a breathing disorder known as apnoea, psychosomatic disorders such as tension, or other physical or mental distress.

Insomnia is not harmful if it is only occasional; the body is readily restored by a few hours of extra sleep. If, however, it is regular or frequent, it may have injurious effects on the other systems and functions of the body.

Treatment of mild insomnia may involve simple improvement of sleeping conditions or such traditional remedies as warm baths, milk, or systematic relaxation.

Severe or chronic insomnia may necessitate the temporary use of barbiturates or tranquilizing drugs, hypnosis, or, because many chronic cases are psychogenic, psychotherapy; apnoea and its associated insomnia may be treated surgically.

The prolonged use of drugs as a relief from frequent or recurring insomnia can have harmful effects. The body tends to build up a tolerance to the soporific qualities of the drug, necessitating more potent dosage; habitual use can lead to addiction.

POST-TRAUMATIC STRESS DISORDER

In this condition symptoms develop in an individual after he has experienced a psychologically traumatic event. The traumatic events can include serious automobile accidents, rape or assault, military combat, torture, incarceration in a concentration or death camp, and such natural disasters as floods, fires, or earthquakes.

A feature of this condition is the person's re-experiencing of the traumatic event in nightmares and in intrusive daytime fantasies. Sometimes an insignificant event, like a knock at the door, will precipitate a sudden terrifying recollection and an exaggerated startle response.

Other symptoms include emotional numbing, a diminished ability to enjoy activities or relationships that were previously pleasurable, and difficulty with sleeping. Long-term symptoms of distress, marital and family problems, difficulties at work, and the abuse of alcohol and other drugs are characteristic impairments caused by the disorder.

The marked emotional symptoms may persist long after the traumatic event actually occurs. Some persons are more liable than others to develop the disorder, depending on personality traits, previous psychological disturbances, age, and genetic predisposition. Psychotherapy is the basic approach used in treating this disorder.

PYROMANIA

Pyromania is the impulse-control disorder characterized by the recurrent compulsion to set fires. The term refers only to the setting of fires for sexual or other gratification provided by the fire itself, not to arson for profit or revenge. Pyromania is usually a symptom of underlying psychopathology, often associated with aggressive behaviours.

Sigmund Freud, the founder of psychoanalysis, noted that the majority of pyromaniacs are males with a history of bed-wetting and suggested that pyromania is one of many disorders brought on by the denial of instinctual drives, in this case a male desire to control fire by urination. Later psychoanalysts found his explanation too simplistic. Among other suggested causes of pyromania are the feelings of rejection and the wish for the return of an absent father.

Pyromania usually first surfaces in childhood and only a small percentage of adult fire-setters actually suffer from the disorder. Pyromaniacs fighting an urge to set fires experience increasing tension that can only be relieved by giving in; after repeated failures to control the impulse, they may cease resistance to avoid this tension. The disorder may be treated by family-centred psychotherapy and by antidepressant drugs.

9. CLARIFICATION OF DISORDERS

MINOR AFFECTIVE DISORDERS

This is a less severe manifestation of the manic-depressive syndrome, in which the mood swings are present but not as extreme, is termed cyclothymic disorder.

This illness is better considered a personality disorder of affective type; the prevailing mood swings are established in adolescence and continue throughout adult life.

DYSTHMIC DISORDER

Dysthmic disorder, or depressive neurosis, may occur on its own, but it more commonly appears along with other neurotic symptoms such as anxiety, phobia, and hypochondriasis.

Where there are clear external grounds for a person's unhappiness, a dysthymic disorder is considered to be present when the depressed mood is disproportionately severe or prolonged in regard to the distressing experience, when there is a preoccupation with the precipitating situation, when the depression continues even after removal of the provocation, and when it impairs the individual's ability to cope with the specific stress.

At any time, depressive symptoms may be found in one-sixth of the population, more commonly in women than men. Social factors are important etiologically, as evidenced in the high rates of depression found in urban women living without a male cohabitant, having three or more children, and lacking employment outside the home.

Loss of self-esteem, feelings of helplessness and hopelessness, and losses of various types of "loved objects" are also seen as important causes of minor depression.

The course and severity of dysthymic disorder is extremely variable-- from a few weeks or months to several decades and from the mild impairment of social functioning to almost total incapacitation. Psychotherapy is the treatment of choice, although antidepressant medication may prove beneficial.

HYPOCHONDRIASIS

This mental disorder is characterized by an excessive preoccupation with one's own health and a tendency to fear or believe that one has a serious disease based on the presence of insignificant physical signs or symptoms.

The hypochondriac may become convinced that he is ill even though such physical signs are completely absent, or he may exaggerate the medical significance of minor aches and pains, becoming morbidly and obsessively preoccupied with the thought of a life-threatening illness.

His fears usually persist even after a thorough examination by a physician has established that no physical abnormality exists, and the doctor's reassurances have only a slight or temporary effect on such a person's apprehensions.

The typical hypochondriac does not become delusional about his health, however; *i.e.,* he remains able to consider or admit the possibility that his fears are unfounded. Despite this, many hypochondriacs will go from doctor to doctor in their effort to enlist medical resources to deal with the imagined illness.

A hypochondriac's concern may focus on a supposed disease or diseases of the heart, gastrointestinal tract, or genital structures. The person may become fixated on one imagined disorder, or he may fasten on different ones as time goes by according to his own readings of medical literature, current public health concerns and fads, and so on.

Hypochondriasis may exist by itself or it may be a secondary syndrome that occurs along with another mental disorder.

Hypochondriasis usually firsts manifests itself in young adulthood and is equally common among males and females. In some cases it seems to represent a psychological coping mechanism that the person resorts to in order to deal with stressful life situations.

Psychotherapy can be helpful in determining an underlying emotional cause, and behaviour therapy may also be helpful.

AMNESIA

Amnesia means loss of memory as a result of brain injury or deterioration, shock, fatigue, senility, drug use, alcoholism, anaesthesia, illness, or psychoneurotic reaction. Amnesia may be anterograde, in which events following causative trauma or disease are forgotten; or retrograde, in which events preceding the causative event are forgotten.

The condition can often be traced to some severe emotional shock, in which case personal memories (*e.g.,* identity), rather than less personal material (*e.g.,* language skills), are affected. Such amnesia seems to represent a psychoneurotic escape from or denial of memories that might cause anxiety and is thus an example of repression, or motivated forgetting.

These memories are not actually lost, since they can generally be recovered through psychotherapy or after the amnesic state has ended.

Occasionally amnesia may last for weeks, months, or even years, during which time the person may begin an entirely new life pattern. Such protracted reactions are called fugue states. When recovered, the person is usually able to remember events that occurred prior to onset, but events of the fugue period are forgotten.

Posthypnotic amnesia, the forgetting of most or all events that occur while under hypnosis in response to a suggestion by the hypnotist, has long been regarded as a sign of deep hypnosis.

The common difficulty of remembering childhood experiences is sometimes referred to as childhood amnesia, or infantile amnesia.

HYSTERICAL AMNESIA

Hysterical amnesia is of two main types. One involves the failure to recall particular past events or those falling within a particular period of the patient's life. This is essentially retrograde amnesia but it does not appear to depend upon an actual brain disorder, past or present. In the second type there is failure to register--and, accordingly, later to recollect--current events in the patient's ongoing life.

This is essentially anterograde amnesia and, as an ostensibly psychogenic phenomenon, would appear to be rather rare and almost always encountered in cases in which there has been a pre-existing amnesia of organic origin.

Rarely, amnesia appears to cover the patient's entire life, extending even to his own identity and all particulars of his whereabouts and circumstances. Although most dramatic, such cases are extremely rare and seldom wholly convincing. They usually clear up with relative rapidity, with or without psychotherapy.

Hysterical amnesia differs from organic amnesia in important respects. As a rule it is sharply bounded, relating only to particular memories, or groups of memories, often of direct or indirect emotional significance.

It is also usually motivated in that it can be understood in terms of the patient's needs or conflicts; *e.g.*, the need to seek financial compensation after a road accident causing a mild head injury or to escape the memory of an exceptionally distressing or frightening event.

Hysterical amnesia also may extend to basic school knowledge, such as spelling or arithmetic, which is never seen in organic amnesia unless there is concomitant aphasia or a very advanced state of dementia.

A most distinctive feature of hysterical amnesia is that it can almost always be relieved by such procedures as hypnosis. Although distinguishing organic from psychogenic amnesia is not always easy, it can usually be achieved on the basis of such criteria, especially when there is no reason to suspect actual brain damage.

TIC

Tic is a sudden rapid, recurring contraction in a muscle or group of muscles, occurring more often in the upper parts of the body. The movement is always brief, irresistible, and limited to one part of the body. It does not interfere with the use of the part involved and may be halted voluntarily, but only for a time.

If the tic movement becomes ingrained, it is looked upon as a habit; the possessor becomes relatively unaware of its occurrence. Then a tic may be considered to be involuntary. These characteristics of tic help to differentiate it from other involuntary movements, such as spasm or cramp, or the capricious, uninhabitable movements that may occur in the afflictions chorea (St. Vitus' dance) or epilepsy.

While most tics are probably of psychological origin, similar repetitive movements have been observed in physical disorders, as in the late stages of encephalitis. The movements that accompany brain disease may persist for years but tend to cease eventually.

Nervous children between five and twelve years of age are those most likely to have tics, but no age group is immune. The movement appears when the subject is tense, and distraction will reduce it. The sufferer knows that he has a certain control of the movement but feels impelled to go through with it in order to feel better.

According to psychoanalytic interpretation, tics are the involuntary motor expression of emotional activity, and they depend primarily upon the subject's early psychological and sexual development.

Tics frequently involve the face and air passages. The most common tics are a repetitive grimace, blink, sniff, snort, or click in the nose or throat, a twitch, or a shrug.

Tics occur in decreasing frequency from head to foot. Psychotherapy, relaxation training, and biofeedback training have had some success in treating tics.

STRESS

Stress symptoms are found in psychology and biology. This may be any strain or interference that disturbs the functioning of an organism. The human being responds to physical and psychological stress with a combination of psychic and physiological defences.

If the stress is too powerful, or the defences inadequate, a psychosomatic or other mental disorder may result.

Stress is an unavoidable effect of living and is an especially complex phenomenon in modern technological society. There is little doubt that an individual's success or failure in controlling potentially stressful situations can have a profound effect on his ability to function.

The ability to "cope" with stress has figured prominently in psychosomatic research. Researchers have reported a statistical link between coronary heart disease and individuals exhibiting stressful behavioural patterns designated "Type A."

These patterns are reflected in a style of life characterized by impatience and a sense of time urgency, hard-driving competitiveness, and preoccupation with vocational and related deadlines.

Various strategies have been successful in treating psychological and physiological stress. Moderate stress may be relieved by exercise and any type of meditation (*e.g.*, yoga or Oriental meditative forms). Severe stress may require psychotherapy to uncover and work through the underlying causes.

A form of behaviour therapy known as biofeedback enables the patient to become more aware of internal processes and thereby gain some control over bodily reactions to stress. Sometimes, a change of environment or living situation may produce therapeutic results.

PHOBIA

Phobia is an extreme, irrational fear of a specific object or situation. A phobia is classified as a type of anxiety disorder, since anxiety is the chief symptom experienced by the sufferer. Phobias are thought to be learned emotional responses.

It is generally held that phobias occur when fear produced by an original threatening situation is transferred to other similar situations, with the original fear often repressed or forgotten. An excessive, unreasoning fear of water, for example, may be based on a forgotten childhood experience of almost drowning.

The person accordingly tries to avoid that situation in the future, a response that, while reducing anxiety in the short term, reinforces the person's association of the situation with the onset of anxiety.

Behaviour therapy is often successful in overcoming phobias. In such therapy, the phobic person is gradually exposed to the anxiety-provoking object or situation in a controlled manner until he eventually ceases to feel anxiety, having realized that his fearful expectations of the situation remain unfulfilled.

In this way, the strong associative links between the feared situation, the person's experience of anxiety, and his subsequent avoidance of that situation are broken and are replaced by a less-maladaptive set of responses.

Psychotherapy may also be useful in the treatment of phobias.

Although psychiatrists classify phobias as a single type of anxiety disorder, hundreds of words have been coined to specify the nature of the fear by prefixing "phobia" with the Greek word for the object feared.

Among the more common examples are acrophobia, fear of high places; claustrophobia, fear of closed places; nyctophobia, fear of the dark; ochlophobia, fear of crowds; xenophobia, fear of strangers; and zoophobia, fear of animals.

Agoraphobia, the fear of being in open or public places, is a particularly crippling illness that may prevent its victims from even leaving home. School phobia may afflict schoolchildren who are overly attached to a parent.

PHOBIC DISORDER OR NEUROSIS

Phobias are neurotic states accompanied by intense dread of certain objects or situations that would not normally have such an effect. This type of anxiety is associated with a strong desire to avoid the dreaded object or situation.

About six per 1,000 of the population suffer from a phobic disorder. There is a tendency for phobic symptoms, whatever their nature, to persist for many years unless treated, and the avoidance behaviour they produce can seriously limit the affected individual's movements and his social or occupational functioning.

People can have phobias about many different kinds of objects or situations, but three main divisions of phobic syndromes are made by the simple phobia, agoraphobia, and social phobia.

Individuals with simple phobias may intensely fear a specific object or situation, for example, cats or thunderstorms; they have anxious thoughts upon anticipating contact with an object or event, for instance, upon hearing the weather forecast, and they try to avoid the object, as in staying indoors in order not to encounter a cat.

Typically, agoraphobic patients have an intense fear of being alone in or being unable to escape from a public place or some other setting outside the home, such as a crowded bus or a supermarket. A social phobia is present when the individual has extreme anxiety in a social situation where he is under the scrutiny of others, such as eating in a restaurant or speaking at a meeting.

The treatment of phobic disorders is best approached by the use of behavioural therapy; dynamic psychotherapy and anti-anxiety drugs may be effective in some cases.

ANOREXIA NERVOSA

Anorexia nervosa is the opposite of obesity. It seems to occur only in societies where adequate food is taken for granted and where people may feel anxiety about avoiding obesity. In affluent communities many teenage girls and young women diet to stay slim or become slim, though their attempts may not be successful.

The young woman with anorexia nervosa, on the other hand, does not talk about dieting but succeeds in losing the weight that the others talk about losing. She does have an appetite but it is strongly suppressed.

By rigid control of her eating she avoids foods that she understands to be fattening. She has a phobia of being fat and often has a distorted body image, seeing herself in the mirror as fatter than she really is.

Amenorrhea (cessation of menstruation) is an early characteristic. Before the loss of weight she was often a model of good behaviour, conformism, and achievement, though this probably concealed a sense of ineffectiveness and self-doubt.

Up to one in 100 middle-class women from 15 to 25 years may be affected. Some women not only abstain; they have learned to induce vomiting or purging and may have eating binges in between. When habitual, this behaviour is called bulimia nervosa.

The physical effects of a young woman starving herself down to 90 pounds (45 kilograms) or less differ in several ways from those of famine. The young woman with anorexia nervosa usually eats adequate protein and micronutrients; she is restless and overactive. She denies that she is too thin or that she is not eating enough.

Treatment is easiest if the condition is recognized at an early stage. It is best managed by a specialized team, most often a psychiatrist with experience in anorexia nervosa, working with a dietician. The principles of treatment are behavioural therapy, supervised eating, and supportive psychotherapy.

OBSESSIVE-COMPULSIVE DISORDER OR NEUROSIS

In this condition an individual experiences obsessions or compulsions or both. Obsessions are recurring words, thoughts, ideas, or images that, rather than being experienced as voluntarily produced, seem to invade a person's consciousness despite his attempts to ignore, control, or suppress them.

The obsessional thought or topic is perceived by the sufferer as inappropriate or senseless; the idea is recognized both as alien to his nature and yet as coming from inside himself. An obsession can take the form of a recurrent and vivid fantasy that is often obscene, disgusting, repugnant, or senseless.

The patient with obsessional ruminations holds endless debates over mundane matters inside his head; *e.g.,* "Did I forget to lock the front door behind me?"

Obsessions in turn are frequently linked to compulsions. These are urges or impulses to perform repetitive acts that are apparently meaningless, unnecessary, stereotyped, or ritualistic.

The compulsive person knows that the act to be performed is meaningless or unnecessary, but his failure or refusal to perform it brings on a mounting tension or anxiety that is temporarily relieved once the act is performed.

Obsessional ruminations thus directly produce compulsive behaviour; *e.g.,* repeatedly checking and relocking an already secure front door.

Most compulsive acts have a simple, ritualistic character and can involve checking, touching, hand-washing, or the repetition of particular words or phrases.

DRUGS

Psychotherapy and behavioural therapy are selectively successful in treating obsessive-compulsive disorders, depending on the individual patient.

The drug clomipramine has proved to be notably effective in reducing or even eliminating the symptoms in a large proportion of patients tested.

PREGNANCY

In this case one looks for conditions that may be mistaken for pregnancy, or other conditions which may confuse the diagnosis of pregnancy.

Absence of menstruation can be caused by chronic illness, by emotional or endocrine disturbances, by fear of pregnancy, or by a desire to be pregnant. Nausea and vomiting may be of gastrointestinal or psychic origin. Tenderness of the breasts can be due to a hormonal disturbance.

Any condition that causes pelvic congestion, such as a pelvic tumour, may cause duskiness of the genital tissues. At times a soft tumour of the uterus may simulate a pregnancy.

The question of pregnancy may be raised if the woman does not menstruate regularly; the absence of other symptoms and signs of gestation indicates that she is not pregnant.

There are rare ovarian and uterine tumours that produce false-positive pregnancy tests. It may be difficult for the physician to exclude pregnancy on the basis of an examination if the uterus is tipped back and difficult to feel, or if it is enlarged by a tumour within it.

If other signs of pregnancy are absent, however, and the tests for pregnancy are negative, pregnancy can most likely be ruled out.

Childless women who greatly desire a baby sometimes suffer from false or spurious pregnancy (pseudocyesis). They stop menstruating, have morning nausea, "feel life," and have abdominal enlargement caused by fat and intestinal gas.

At "term" they may have "labour pains." Signs of pregnancy are absent. Treatment is by psychotherapy.

Menopausal women often fear pregnancy when their periods stop; information that they show no signs of pregnancy usually reassures them.

Retained uterine secretions of bloody or watery fluid, caught above a blocked mouth of the uterus (cervix), prevent menstruation, cause softening and enlargement of the uterus, and may cause the patient to wonder whether she is pregnant. There are no other signs of pregnancy, and the hard cervix, closed by scar tissue, explains the problem.

FUNCTIONAL PSYCHOSES

Schizophrenia is the most common and the most potentially severe and disabling of the psychoses. Schizophrenia is characterized by a withdrawal from reality, delusions and hallucinations, a loosening and incoherence of a person's thought processes, and deficiencies in feeling appropriate or normal emotions.

Other symptoms may include apathy, reduced drive and initiative, inability to feel any emotion whatsoever, and a preoccupation with silly or bizarre fantasies. The symptoms of schizophrenia typically first manifest themselves during the teen years or early adult life.

The course of the disease is variable: some schizophrenics suffer one acute episode and then permanently recover; others suffer from repeated episodes with periods of remission in between; and still others become chronically psychotic and must be permanently hospitalized.

Despite prolonged research, the cause or causes of schizophrenia remain largely unknown. It is clear, however, that there is a genetic predisposition to the disease that is inherited. Thus the children of schizophrenic parents stand a greatly increased chance of themselves becoming schizophrenic.

The symptoms of schizophrenia can be treated, but not cured, with such antipsychotic drugs as chlorpromazine and other phenothiazine drugs and by haloperidol. Psychotherapy may be useful in alleviating distress and helping the patient to cope with the effects of his illness.

The affective psychoses, which are also known as major affective disorders, consist of states of extreme and prolonged depression, extreme mania, or alternating cycles of both of these mood abnormalities. Depression is a sad, hopeless, pessimistic feeling that can cause listlessness; loss of pleasure in one's surroundings, loved ones, and activities; fatigue; slowness of thought and action; insomnia; and reduced appetite.

Mania is a state of undue and prolonged excitement that is evinced by accelerated, loud, and voluble speech; heightened enthusiasm, confidence, and optimism; rapid and disconnected ideas and associations; rapid or continuous motor activity; impulsive, gregarious, and overbearing behaviour; heightened irritability; and a reduced need for sleep.

When depression and mania alternate cyclically or otherwise appear at different times in the same patient, the person is termed to be suffering from a manic-depressive psychosis.

Manic-depressive patients also frequently suffer from delusions, hallucinations, or other overtly psychotic symptoms. Manic depression often first manifests itself around age 30, and the disease is a long-lasting one.

Many such patients can be treated by long-term maintenance on lithium carbonate, which reduces and prevents the attacks of mania and depression.

PSYCHOTIC DEPRESSION

Depression alone can be psychotic if it is severe and disabling enough, and particularly if it is accompanied by delusions, hallucinations, or paranoia. Depression can be effectively treated by a variety of antidepressant drugs, including the tricyclic antidepressants and the monoamine oxidase inhibitors.

Electroconvulsive (shock) therapy is useful in some cases, and psychotherapy and behavioural therapy may also be effective. Mania and many cases of depression are believed to be caused by deficiencies or excesses of certain neurotransmitters in the brain. (Neurotransmitters are chemicals that play key roles in the transmission of nerve impulses.)

LEARNED HELPLESSNESS

This is a theory of depression formulated by the US psychologist Martin Seligman in his book *Helplessness* (1975) and later developed into an attributional theory (see social understanding). Learned helplessness occurs in response to prolonged or repeated inescapable punishment of some sort, the key symptom being a passive, listless, or defeatist response to simple problems.

Bad events are believed to be caused by factors--such as stupidity, or weakness of will--which are 'internal' (an aspect of oneself), 'stable' (will remain), and 'global' (affect other areas of one's life). Good events in contrast are believed to have external, transient, and specific causes, such as momentary good luck.

The concept is therefore related to other locus of control theories. Psychotherapy may provide effective treatment.

PARANOIA

Paranoia is a special syndrome that can be a feature of schizophrenia (paranoid schizophrenia) and manic-depressive psychosis or can exist by itself. A person suffering from paranoia thinks or believes that other people are plotting or trying to harm, harass, or persecute him in some way. The paranoiac exaggerates trivial incidents in everyday life into menacing or threatening situations and cannot rid himself of his suspicions and apprehensions. Paranoid syndromes can sometimes be treated or alleviated by antipsychotic drugs.

The functional psychoses are difficult to treat, and drug treatments are the most common and successful approach. Psychoanalysis and other psychotherapies, which are based on developing a patient's insight into his presumed underlying emotional conflicts, are difficult to apply to psychotic patients.

PSYCHOSEXUAL DYSFUNCTION

Psychosexual dysfunction is the inability of a person to experience sexual arousal or to achieve sexual satisfaction under appropriate circumstances, as a result of either physical disorder or, more commonly, psychological problems.

The most common forms of sexual dysfunction have traditionally been classified as impotence (inability of a man to achieve or maintain penile erection) and frigidity (inability of a woman to achieve arousal or orgasm during sexual intercourse). Because these terms--impotence and frigidity--have developed pejorative and misleading connotations, they are no longer used as scientific classifications, having been superseded by more specific terms; however, both terms remain in common usage, with a variety of meanings and associations (*see* frigidity; impotence).

Sexual dysfunctions recognized by professional therapists include hyposexuality (or inhibited sexual excitement), in which sexual arousal can be achieved only with great difficulty; anorgasmia, in which a woman has a recurrent and persistent inability to achieve orgasm despite normal sexual stimulation; vaginismus, in which the woman's vaginal muscles contract strongly during intercourse, making coitus difficult or impossible; dyspareunia, in which a woman experiences significant pain during attempts at intercourse; erectile impotence, in which a man cannot sustain an erection; ejaculatory impotence (or inhibited male orgasm), in which a man cannot achieve orgasm in the woman's vagina, although he can sustain an erection and may reach orgasm by other methods; and premature ejaculation, in which the man ejaculates before or immediately after entering the vagina.

In most cases, each of these dysfunctions reflects the individual's anxiety or other negative feelings about the sex act or partner, although emotional conflicts outside the sexual relationship itself can also produce failures of sexual function. Appropriate sex therapy, designed to help the individual relax in his or her sexual role, can often overcome the anxiety and eliminate the dysfunction, although the success of such therapy varies markedly among the various dysfunctions and among individual patients.

When a specific physical condition predisposes to the dysfunction, it must be treated medically; alcoholism and endocrine or neurological disorders are among the common physical causes of sexual dysfunction. Sexual dysfunctions that are secondary to more severe psychological or personality disorders may require specific psychotherapy.

ALCOHOLISM

Psychotherapy in alcoholism encompasses the entire range of modalities applied in treating the psychoneuroses and character disorders, including individual and group techniques. The aim varies from eliminating some underlying cause to effecting just enough shift in the patient's emotional state so that he can function at least temporarily without drinking. Psychoanalysis is rarely tried, having shown little success in alcoholism; analytically oriented therapies are more usual, chiefly with supportive aims.

The only psychological technique developed specifically for alcoholism consists of gaining the patient's recognition and acceptance of his actual condition, which alcoholics often resist. Such acceptance may then be followed by a therapeutic-rehabilitative regimen.

Group therapies are regarded as more effective than individual modalities with alcoholics. These range from instructional lectures and superficial discussions to deep analytic explorations, psychodrama, hypnosis, psychodynamic confrontation, and marathon sessions. Mechanical aids include didactic motion pictures, movies of the patients while intoxicated, and taped records of previous sessions.

Some therapists have experimented, as yet without definitive results, with milieus that reward and reinforce socializing behaviour, hoping thereby to extinguish the de-socializing drinking behaviour. Many institutional programs rely on "total push," subjecting the patient to a bombardment of methods, including drugs, hypnosis, physiotherapies, group sessions, lectures,

Alcoholics Anonymous meetings, and individual psychological and religious counselling, with the hope that each patient will be affected favourably by whatever is most suitable for him. Other institutional programs rely on mere removal from the stressful outside environment, with a period of enforced abstinence. The therapists themselves may be psychoanalysts, psychiatrists, clinical psychologists, pastoral counsellors, social workers, nurses, police or parole officers, or lay counsellors--the latter often former alcoholics with special training.

The places of treatment are as varied as the modalities, ranging from general hospitals to mental hospitals to mental-health outpatient clinics to specialized inpatient sanitariums and specialized alcoholism clinics to jails and penitentiaries to medical and psychiatric private offices, with patients often moving, randomly or systematically, from one milieu to another.

Awareness of the social and environmental elements in alcoholism has led to the development of treatment for spouses and occasionally for whole families, either separately or jointly, in the recognition that "the patient" is not just the alcoholic but the family unit. A new trend in the United States, partly stimulated by court decisions prohibiting the jailing of alcoholics for public intoxication, is the establishment of detoxication centres that provide first aid along with guidance toward more fundamental treatment. But even if adequate programs and facilities for treating alcoholism were available, it is unlikely that they would solve the problem, given the large number of new cases each year. Only preventive public-health programs can eliminate alcoholism and thus far no likely methods of prevention have been devised.

PRE-MENSTRUAL SYMPTOMS (PMS)

This is group of physical and emotional symptoms that occur in women before the onset of menstruation and that are characteristically cyclical in nature. These symptoms generally begin from 7 to 14 days before menstruation and end within 24 hours after menstruation has begun. The medical condition, termed premenstrual syndrome by British physician Katharina Dalton in the 1950s, has only been studied since about the 1930s. In 1985 biologists working in Kenya found that premenstrual female baboons were subject to a similar phenomenon; the premenstrual baboons were observed to withdraw from social contact and to spend a much greater percentage of time up in the trees feeding.

Physical symptoms may include headache, cramps, backache, bloating, constipation or diarrhoea, and a number of related disorders. Emotional symptoms range from irritability, lethargy, and rapid mood swings to hostility, confusion, aggression, and severe depression.

Though they are the major subject of current research, the causes of PMS are not yet established. The most widely accepted theories centre on hormonal changes (the rapid fluctuation of levels of oestrogen and progesterone in the bloodstream), nutritional deficiencies (particularly in regard to the vitamins--notably B vitamins--that affect nerve transmission in the brain), and stress (which has been shown to be a factor in the severity of symptoms). Many researchers suspect that fluctuations of chemical transmitters in the brain are largely responsible.

For purposes of treatment, a chart that records the nature and date of occurrence of an individual's symptoms can aid diagnosis. The major method of treatment for most cases of PMS involves some combination of regular physical exercise, avoidance of stress, and nutritional therapy. Restriction of sodium intake, avoidance of xanthines--found in coffee, tea, chocolate, and cola--and a diet that is high in protein and complex carbohydrates are a few of the dietary measures that can be taken to alleviate or reduce much of the physical discomfort. More severe cases may require drug therapy (including the use of analgesics, diuretics, antidepressants, and sedatives) and psychotherapy to aid stress management.

10. PSYCHIATRY AND MENTAL DISORDERS

DISSOCIATIVE DISORDERS OR HYSTERICAL NEUROSIS

Dissociation is a syndrome in which one or a group of mental processes are split off, or dissociated, from the rest of the psychic apparatus so that their function is lost, altered, or impaired. Dissociative symptoms have often been regarded as the mental counterparts of the physical symptoms displayed in conversion disorders. Since the dissociation may be an unconscious mental attempt to protect the individual from threatening impulses or emotions that are repressed, the conversion into physical symptoms and the dissociation of mental processes can be seen as related defence mechanisms arising in response to emotional conflict.

In dissociative disorders there is a sudden, temporary alteration in the person's consciousness, sense of identity, or motor behaviour. There may be an apparent loss of memory of previous activities or important personal events, with amnesia for the episode itself after recovery. These are rare conditions, and it is important to exclude organic causes.

In psychogenic or hysterical amnesia there is a sudden loss of memory which may appear total; the patient can remember nothing about his previous life or even his name. The amnesia may be localized to a short period of time associated with a traumatic event or it may be selective, affecting the person's recall of some, but not all, of the events during a particular time. In psychogenic fugue, the individual wanders away from his home or place of work and assumes a new identity; he cannot remember his previous identity and upon recovering cannot recall the events that occurred while he was in the fugue state. In many cases the disturbance lasts only a few hours or days and involves only limited travel. Severe stress frequently triggers this disorder.

MULTIPLE PERSONALITY

Multiple personality is a rare and remarkable dissociative disorder in which two or more distinct and independent personalities develop in a single individual. Each of these personalities inhabits the person's conscious awareness to the exclusion of the others at particular times. This disorder frequently arises as a result of traumas suffered during childhood and is best treated by psychotherapy, which seeks to reunite the various personalities into a single, integrated one.

DEPERSONALIZATION

In depersonalization a person feels or perceives his body or self as being unreal, strange, altered in quality, or distant. This state of self-estrangement may take the form of feeling as if one is machinelike, is living in a dream, or is not in control of one's actions. Derealization, or feelings of unreality concerning objects outside one's self, often occurs at the same time. Depersonalization may occur alone in neurotic patients but is more often associated with phobic, anxiety, or depressive symptoms. It most commonly occurs in younger married women and may persist for many years.

Patients find the experience of depersonalization intensely difficult to describe and often fear that people will think them insane. Organic conditions, especially temporal lobe epilepsy, must be excluded before making a diagnosis of neurosis when depersonalization occurs. As with other neurotic syndromes, it is more common to see a mixed picture with many different symptoms than depersonalization alone.

The causes of depersonalization are obscure, and there is no specific treatment for it. When the symptom arises in the context of another psychiatric condition, treatment is aimed at that illness.

SCHIZOPHRENIA

Schizophrenia is one of any of a group of severe mental disorders that have in common such symptoms as hallucinations, delusions, blunted emotions, disordered thinking, and a withdrawal from reality.

Schizophrenics display a wide array of symptoms, but four main types of schizophrenia, differing in their specific symptomatology as follows, are recognized by some authorities:

1. The simple or undifferentiated type of schizophrenic manifests an insidious and gradual reduction in his external relations and interests. His emotions lack depth, his ideation is simple and refers to concrete things, and there is a relative absence of mental activity, a progressive lessening in the use of inner resources, and a retreat to simpler or stereotyped forms of behaviour.

2. The hebephrenic or disorganized type of schizophrenic displays shallow and inappropriate emotional responses, foolish or bizarre behaviour, false beliefs (delusions), and false perceptions (hallucinations).

3. The catatonic type is characterized by striking motor behaviour. The patient may remain in a state of almost complete immobility, often assuming statuesque positions. Mutism (inability to talk), extreme compliance, and absence of almost all voluntary actions are also common. This state of inactivity is at times preceded or interrupted by episodes of excessive motor activity and excitement, generally of an impulsive, unpredictable kind.

4. The paranoid type, which usually arises later in life than the other types, is characterized primarily by delusions of persecution and grandeur combined with unrealistic, illogical thinking, often accompanied by hallucinations.

These different types of schizophrenia are not mutually exclusive, and schizophrenics may display a mixture of symptoms that defy convenient classification. There may also be a mixture of schizophrenic symptoms with those of other psychoses, notably those of the manic-depressive group.

Hallucinations and delusions, although not invariably present, are often a conspicuous symptom in schizophrenia. The most common hallucinations are auditory: the patient hears (nonexistent) voices and believes in their reality. Schizophrenics are subject to a wide variety of delusions, including many that are characteristically bizarre or absurd. One symptom common to most schizophrenics is a loosening in their thought processes; this syndrome manifests itself as disorganized or incoherent thinking, illogical trains of mental association, and unclear or incomprehensible speech.

Schizophrenia crosses all socioeconomic, cultural, and racial boundaries. The lifetime risk of developing the illness has been estimated at about 8 per 1,000. Schizophrenia is the single largest cause of admissions to mental hospitals and accounts for an even larger proportion of the permanent populations of such institutions. The illness usually first manifests itself in the teen years or in early adult life, and its subsequent course is extremely variable. About one-third of all schizophrenic patients make a complete and permanent recovery, one-third have recurring episodes of the illness, and one-third deteriorate into chronic schizophrenia with severe disability.

Various theories of the origin of schizophrenia have centred on anatomical, biochemical, psychological, social, genetic, and environmental causes. Despite much research, no single cause of schizophrenia has been established, or even identified. There is strong evidence that genetic inheritance often plays a role in the disease, and researchers continue to search for the presence of biochemical abnormalities in the brains of people suffering from schizophrenia. Stressful life experiences may trigger the disease's initial onset.

There is no cure for most patients with chronic schizophrenia, but the disease's symptoms can in many cases be effectively treated by antipsychotic drugs given in conjunction with psychotherapy and supportive therapy. *See also* psychosis.

SPEECH DISORDER

Stuttering or stammering is academically known as dysphemia. What is called stuttering in the United States is usually named stammering in Great Britain. While everyone seems to know what stuttering sounds like, experts do not agree about what really causes it. In the age groups after puberty, stuttering is the most frequent and conspicuous type of disturbed speech encountered. This is one reason why among the studies dealing with speech pathology in the world literature those devoted to stuttering are the largest single group.

Despite numerous and intensive studies of the problem, findings and conclusions are far from unanimous. A great number of theories have been proposed to explain the origin and nature of stuttering, which range from the premise that subtle physical disturbances in the nervous system (so-called neurogenic asynchronies) are responsible to the opinion that psychological maladjustment alone is to blame. The report issued by the U.S. Public Health Service (see above Prevalence of speech disorders) concludes that "stuttering remains an enigma while illustrating the type of disorder which does not have either a clean-cut organic cause or a clearly habitual basis."

Research findings indicate (as is the case with many developmental speech disorders, particularly language disability, articulatory disorders, reading disability, and cluttering) that trouble with stuttering affects the male sex at least four times more frequently than the female. Hereditary predisposition has been noted in many studies of large groups of stutterers, with evidence for an inherited tendency found among as many as 40 percent of the stutterers studied. Some experts insist that stuttering is not a single disease entity but that it comprises several types of the disorder with different causes.

According to such views, the familial occurrence of stuttering represents a combination of the stuttering symptom with a cluttering tendency that is inherited. Although imitation of another stutterer may form the basis for acquiring the habit, purely psychological explanations that stress parental attitudes in training their children fail to reveal why many stutterers have siblings (brothers or sisters) with perfectly normal speech.

The treatment of stuttering is difficult and often demands much skill and responsibility on the part of the therapist. The possibility of some specific medical cure seems remote at the present time. Even the most advanced methods of modern psychiatry have failed to produce superior results in treatment. For a time it was hoped that new psychopharmacological drugs (*e.g.*, tranquillizers) might facilitate and accelerate recovery from stuttering, but these efforts have been disappointing thus far.

The typical program of management in this disorder is a strict program of psychotherapy (talking freely with a psychiatrist or psychologist so as to reduce emotional problems) supported by various applications of learning theory or behavioural theory (in retraining the stutterer) and other techniques depending on the therapist's position. It is widely agreed that the patient must acquire a better adjustment to the problems of his life and that he needs to develop a technique for controlling his symptoms and fears. Prognosis (predicted outcome of treatment) thus is held to depend greatly on the patient's motivation and perseverance.

It is interesting to note that experienced investigators no longer aspire to a "cure" of stuttering through an etiologic (causal) approach. Instead of focussing on underlying causes, they aim at making the patient "symptom-free" via symptomatic therapy. Prevention of stuttering may be aided through parent counselling. The normal, immature speech of many children is characterized by various non-fluencies; these include hesitations, syllable repetition, groping for the right word, and vocalizations between words such as "ah-ah." Some misguided parents castigate these normal signs of developing speech with various admonitions and, even worse, try to forbid the non-fluencies by mislabelling them as stuttering. In some children, this parental interference associates normal non-fluency with feelings of insecurity and fear, tending to make the child become a real stutterer.

Much research has been devoted to this probable aetiology for one type of stuttering; its elimination through parental guidance indeed has been reported to help in reducing the number of stutterers.

MEMORY ABNORMALITY

Memory Abnormality is the disorder that affects the ability to remember. Disorders of memory must have been known to the ancients and are mentioned in several early medical texts, but it was not until the closing decades of the 19th century that serious attempts were made to analyze them or to seek their explanation in terms of brain disturbances. Of the early attempts, the most influential was that of a French psychologist, Théodule-Armand Ribot, who, in his *Diseases of Memory* (1881, English translation 1882), endeavoured to account for memory loss as a symptom of progressive brain disease by embracing principles describing the evolution of memory function in the individual, as offered by an English neurologist, John Hughlings Jackson. Ribot wrote:

The progressive destruction of memory follows a logical order--a law. It advances progressively from the unstable to the stable. It begins with the most recent recollections, which, being lightly impressed upon the nervous elements, rarely repeated and consequently having no permanent associations, represent organization in its feeblest form. It ends with the sensorial, instinctive memory, which, having become a permanent and integral part of the organism, represents organization in its most highly developed stage.

The statement, amounting to Ribot's "law" of regression (or progressive destruction) of memory, enjoyed a considerable vogue and is not without contemporary influence. The notion has been applied with some success to phenomena as diverse as the breakdown of memory for language in a disorder called aphasia and the gradual return of memory after brain concussion. It also helped to strengthen the belief that the neural basis of memory undergoes progressive strengthening or consolidation as a function of time. Yet students of retrograde amnesia (loss of memory for relatively old events) agree that Ribot's principle admits of many exceptions. In recovery from concussion of the brain, for example, the most recent memories are not always the first to return. It has proved difficult, moreover, to disentangle the effects of passage of time from those of rehearsal or repetition on memory.

A Russian psychiatrist, Sergey Sergeyevich Korsakov (Korsakoff), may have been the first to recognize that amnesia need not necessarily be associated with dementia (or loss of the ability to reason), as Ribot and many others had supposed. Korsakov described severe but relatively specific amnesia for recent and current events among alcoholics who showed no obvious evidence of shortcomings in intelligence and judgment. This disturbance, now called the Korsakoff syndrome, has been reported for a variety of brain disorders aside from alcoholism and appears to result from damage in a relatively localized part of the brain.

The neurological approach may be combined with evidence of psychopathology to enrich understanding of memory function. Thus, a French neurologist, Pierre Janet, described amnesia sufferers who were apparently very similar to those observed by Korsakov but who gave no evidence of underlying brain disease. Janet also studied people who had lost memory of extensive periods in the past, also without evidence of organic disorder. He was led to regard these amnesias as hysterical, explaining them in terms of dissociation: a selective loss of access to specific memory data that seem to hold some degree of emotional significance. In his experience, reconnection of dissociated memories could as a rule be brought about by suggestion while the sufferer was under hypnosis.

Freud regarded hysterical amnesia as arising from a protective activity or defence mechanism against unpleasant recollections; he came to call this sort of forgetting repression, and he later invoked it to account for the typical inability of adults to recollect their earliest years (infantile amnesia). He held that all forms of psychogenic (not demonstrably organic) amnesia eventually could resolve after prolonged sessions of talking (psychotherapy) and that hypnosis was neither essential nor necessarily in the amnesiac's best interest. Nevertheless, hypnosis (sometimes induced with the aid of drugs) has been widely used in the treatment of hysterical amnesia, particularly in time of war when only limited time is available.

PAIN

Because pain has both physiological and psychological components, attempts at relief should address both aspects. Helping a patient cope with a painful condition can reduce anxiety, which may lessen the amount of medication needed to alleviate the pain. Acute pain is generally the easiest to control, medication and rest being effective treatments.

Some pain, however, may defy treatment and persist for years. This chronic pain can be compounded by the psychological effects of hopelessness and anxiety.

Opiates are the most potent pain-relieving drugs and are used to treat cases of severe pain. Opium, the dried juice of the opium poppy (*Papaver somniferum*), is one of the oldest and best-known analgesics. Morphine, a powerful opiate, is an extremely effective analgesic.

These narcotic alkaloids mimic the endorphins by binding to their receptors and blocking or reducing the activation of pain neurons.

Use of these narcotics must be monitored not only because opiates are addictive substances but also because the patient can develop a tolerance to them and may require progressively greater doses to achieve the desired level of pain relief.

Significant side effects such as depression and nausea also limit the usefulness of opiates.

Consequently, these narcotics are not prescribed for long-term therapy. They are used to ameliorate pain after surgery and to treat patients with terminal illnesses such as cancer.

In spite of the dangers involved with these narcotics, it has become common to allow the patient to control the amount and frequency of administration of medication with intravenous dispensers.

The rationale behind this strategy is that patients are the highest authority regarding their pain and should therefore be in control of managing it.

Studies indicate that when this method is employed it is not abused; in fact some reports show that less medication is used than would have been prescribed.

Extracts of the bark of the willow tree contain the active ingredient salicin and have been used since antiquity to relieve pain. The modern non-narcotic analgesic salicylates, such as aspirin (acetylsalicylic acid), and salicylate-like drugs, such as acetaminophen, are less potent than the opiates but are non-addictive. They are used to reduce pain resulting from inflammation.

The mode of action of these compounds differs from that of the opiates. They block the body's conversion of arachidonic acid (a cyclic fatty acid) to prostaglandins, which enhance sensitivity to pain.

Psychotropic drugs are used to treat pain that is thought to result from psychological causes alone. These antidepressants and tranquilizers have no effect themselves on the neurophysiologic cause of pain but instead affect the patient's emotional state.

They act by reducing anxiety and altering the patient's perception of the pain. Pain seems to be alleviated in a similar manner by hypnosis, placebos, and psychotherapy. The psychological expectation of relief is itself a potent pain reliever.

Specific nerves can be blocked in cases in which pain is restricted to an area that has few sensory nerves. Phenol is a neurolytic that permanently destroys nerves; lidocaine can be used for temporary relief. Surgical severing of nerves is a radical treatment that is only resorted to infrequently because it can produce serious side effects such as motor loss or delocalised pain.

Some pain can be treated by electrical stimulation. Electrodes are placed on the skin above the painful area, and mild electrical currents are applied. The stimulation of additional peripheral nerve endings has an inhibitory effect on the nerve fibres generating the pain.

This transcutaneous neural stimulation is based on the same process described earlier that allows pain to be inhibited by rubbing the painful area. Acupuncture, compresses, and heat treatment probably operate by the same mechanism.

Chronic pain, defined generally as that which has persisted for at least six months, presents the greatest challenge in pain management. The unrelieved discomfort can significantly alter the life of the sufferer, leading to secondary psychological complications; hypochondriasis and depression are common, as are sleep disturbances, loss of appetite, and feelings of helplessness.

In spite of these negative effects, the behaviour associated with pain can become a habit; the increased attention, sympathy, and support that the sufferer receives reinforce the patient's behaviour, prolonging the pain. This type of learned response illustrates the malleable nature of pain, which can be conditioned by anxiety and fear but is also susceptible to the psychological benefits derived from the disability.

Pain clinics offer a multidisciplinary approach to chronic pain treatment. A distinction is first made between pain behaviour that is a direct response to a noxious stimulus and that which is learned.

If a battery of pain relief methods have been employed with little success, attempts are made to deemphasize reliance on medication and to teach the patient how to live with the pain.

11. PERSONAL AND SOCIAL PROBLEMS

SEXUAL PROBLEMS

Sexual problems may be classified as physiological, psychological, and social in origin. Any given problem may involve all three categories; a physiological problem, for example, will produce psychological effects, and these may result in some social maladjustment.

Physiological problems of a specifically sexual nature are rather few. Only a small minority of people suffer from diseases of or deficient development of the genitalia or that part of the neurophysiology governing sexual response. Many people, however, experience at some time sexual problems that are by-products of other pathologies or injuries.

Vaginal infections, for example, retroverted uteri, prostatitis, adrenal tumours, diabetes, senile changes of the vagina, and cardiovascular conditions may cause disturbance of the sexual life. In brief, anything that seriously interferes with normal bodily functioning generally causes some degree of sexual trouble.

Fortunately, the great majority of physiological sexual problems are solved through medication or surgery. Generally, only those problems involving damage to the nervous system defy therapy.

Psychological problems constitute by far the largest category. They are not only the product of socially induced inhibitions, maladaptive attitudes, and ignorance but also of sexual myths held by society. An example of the latter is the idea that good, mature sex must involve rapid erection, protracted coitus, and simultaneous orgasm.

Magazines, marriage books, and general sexual folklore reinforce these demanding ideals, which cannot always be met and hence give rise to anxiety, guilt, and feelings of inadequacy.

Premature ejaculation is a common problem, especially for young males. Sometimes this is not the consequence of any psychological problem but the natural result of excessive tension in a male who has been sexually deprived. In such cases, more frequent coitus solves the problem. Premature ejaculation is difficult to define.

The best definition is that offered by the American sexologists, William Howell Masters and Virginia Eshelman Johnson, who say that a male suffers from premature ejaculation if he cannot delay ejaculation long enough to induce orgasm in a sexually normal female at least half the time.

This generally means that vaginal penetration with some movement (although not continuous) must be maintained for more than one minute. The average American male ejaculates in two or three minutes after vaginal penetration, a coital duration sufficient to cause orgasm in most females the majority of the time.

Various methods of preventing premature ejaculation have been tried. One is for the male to excite the female more during the foreplay so that she reaches orgasm more rapidly after penetration, but this technique often excites the male as well and defeats its purpose.

Another common method is for the male to think of nonsexual matters, which may prove effective but reduces his pleasure. The most effective therapy is that advocated by Masters and Johnson in which the female brings the male nearly to orgasm and then prevents the male's orgasm by briefly compressing the penis between her fingers just below the head of the penis.

The couple come to realize that premature ejaculation can thus be easily prevented, their anxiety disappears, and ultimately they can achieve normal coitus without resorting to this squeeze technique.

Erectile impotence is almost always of psychological origin in males under 40; in older males physical causes are more often involved. Fear of being impotent frequently causes impotence, and, in many cases, the afflicted male is simply caught up in a self-perpetuating problem that can be solved only by achieving a successful act of coitus.

In other cases, the impotence may be the result of disinterest in the sexual partner, fatigue, distraction because of nonsexual worries, intoxication, or other causes--such occasional impotency is common and requires no therapy.

Some males, however, are chronically impotent and require psychotherapy or behaviour therapy. Such impotency is thought to be the result of deep-seated causal factors such as unconscious feelings of hostility, fear, inadequacy, or guilt. Primary impotence, the inability to ever have achieved erection sufficient for coitus, is more difficult to treat than the far more common secondary impotence, which is impotence in a male who was formerly potent.

Ejaculatory impotence, the inability to ejaculate in coitus, is quite rare and is almost always of psychogenic origin. It seems associated with ideas of contamination or with memories of traumatic experiences. Occasional ejaculatory inability may be expected in older men or in any male who has exceeded his sexual capacity.

Vaginismus is a powerful spasm of the pelvic musculature constricting the vagina so that penetration is painful or impossible. It seems wholly due to anti-sexual conditioning or psychological trauma and serves as an unconscious defence against coitus. It is treated by psychotherapy and by gradually dilating the vagina with increasingly large cylinders.

Dyspareunia, painful coitus, is generally physical rather than psychological. It is mentioned here only because some inexperienced females fear they cannot accommodate a penis without being painfully stretched. This is a needless fear since the vagina is not only highly elastic but enlarges with sexual arousal, so that even a small female can, if aroused, easily receive an exceptionally large penis.

Disparity in sexual desire constitutes the most common sexual problem. It is to some extent inescapable, since differences in the strength of the sexual impulse and the ability to respond are based on neuro-physiological differences.

Much disparity, however, is the result of inhibition or of one person having been subjected to more sexual stimuli during the day than the other. The partner who has been seeing sexually attractive persons periodically during the day and who may have had an opportunity to relax on the way back from the office or store is naturally more interested in coitus than the partner who has not been exposed to sexual stimuli.

Another cause of disparity is a difference in viewpoint. Perhaps one person anticipates coitus as a palliative to compensate for the trials and tribulations of life, whereas another may be interested in sex only if the preceding hours have been reasonably problem-free and happy.

Even in cases of neuro-physiological differences in sex drive, the less-motivated partner can be trained to a higher level of interest, since most humans operate well below their sexual capacities.

Psychological fatigue, a growing disinterest in sexual behaviour with a particular partner, sometimes constitutes a problem. Humans are subject to monotony, and coitus may become routine or even a chore. Lessening frequencies of marital coitus are more often the result of this than of age.

The solution lies in varying the time, the setting, and in breaking away from habitual techniques and positions.

Preferences for or antipathies toward particular positions, techniques, or times frequently cause trouble. One partner may desire mouth-genital contact or anal stimulation that the other partner finds disagreeable or perverse. Some wish to have coitus in the light, others insist upon darkness; some prefer morning, others evening. The possibilities for disagreement are legion.

Even if disagreements stemming from needless inhibition are overcome, there still remain disparities in preference, and these should be met by the philosophy that, by giving pleasure to another, one obtains pleasure. Needless to say, no partner should insist upon that which is abhorrent to the other after the latter has made honest attempts to co-operate.

Lack of female orgasm, anorgasmy, is a very frequent problem. One should differentiate between females who become sexually aroused but do not reach orgasm and those who do not become aroused. Only the latter merit the label frigid. It is common for females not to achieve orgasm during the first weeks or months of coital activity. It is almost as though many females must learn how to have orgasm, for after having had one they respond with increasing frequency.

In some cases, the female initially has no idea how to copulate effectively and simply lies passive, expecting the male to bring her to orgasm. Other females resist orgasm because the feeling of being swept away and losing control is frightening.

In most cases, however, anorgasmy is simply the result of years of inhibition--having been trained since childhood to avoid yielding to the sexual impulse; it is difficult to metamorphose into a responsive and orgasmic being. In the final analysis, anorgasmy is psychological in origin; few, if any, females lack the neurophysiology necessary for orgasm, and anthropology shows that in sexually permissive societies virtually all females have little difficulty in attaining orgasm in coitus.

Anorgasmy is treated by removing inhibitions, by teaching coital techniques, and by inducing orgasm through non-coital methods. The effective therapist should also impress upon the female that not reaching orgasm is no sign of failure or inadequacy on her part or her partner's and that sexual activity is very pleasurable to both, even if orgasm does not ensue.

Indeed, some females derive great pleasure and satisfaction without orgasm, a fact that should be made known to anxious male partners. Too great a concern over orgasm defeats itself. As Kinsey once pointed out, thinking is the enemy of sexual pleasure, and a female can scarcely have orgasm if she is worrying about whether she will attain it or not and if she senses that her partner is mentally turning the pages of a marriage manual.

Lastly, sexual problems are often perpetuated by the inability of the partners to communicate freely their feelings to one another. There is a curious and unfortunate reticence about informing one's partner as to what does or does not contribute to one's pleasure. The partner must function on a trial-and-error basis, ever on the alert for signs indicating the efficacy of his or her efforts.

This muteness is even more pronounced when it comes to an individual making suggestions to the partner. Many persons feel that a suggestion or request would be interpreted by the partner that he or she had been inept or at least remiss. As with any other problems, sexual problems can be overcome or ameliorated only if the individuals concerned communicate effectively.

12. PHILOSOPHY OF MIND

ADEQUACY AS A CRITERION OF THE MENTAL

The question now arises of how adequate subjective experience is as a criterion of the mental--whether, though it is obviously a sufficient condition for something to be mental, it is a sufficient condition for something to have a mind.

The Scotsman David Hume, an 18th-century philosophical Sceptic and historian, once asked whether a creature that had but one state of consciousness could be said to have a mind and concluded that it could not. In his view, it takes, at the very least, a number of states of consciousness linked by memory before one would say that the creature has a mind; and it may be that there has to be a certain level of complexity in the nature and relation of the conscious states for there to be a mind.

It is doubtful, however, whether consciousness is a necessary condition for the mental. Before Sigmund Freud, it would have been widely agreed that the notion of unconscious mental phenomena was logically impossible--a contradiction in the very terms. That view had one important exception, however: Gottfried Wilhelm Leibniz, a 17th-century Rationalist and mathematician held that there are *petites perceptions* of which the subject is unconscious. They are so slight, so similar to others, so familiar, or in such a crowd of other perceptions, that the subject is unaware of them at the time.

One of the examples that Leibniz cited is the person who is unaware of the roar of the waterfall or the rumble of the mill if he has lived nearby for some time. Leibniz seemed to have had in mind what modern psychologists call "subliminal" perceptions, viz., those below the threshold of awareness but still capable of leaving some effects on the mind. But Leibniz confined unconscious states to perceptions; he would not have allowed unconscious beliefs, desires, emotions, or judgments.

It was Freud's great contribution to have discovered a range of phenomena of which the patient was unconscious but which were very much like typically mental phenomena, especially in their behavioural manifestations. In the light of such similarities, it was plausible to extend the concept of the mental to include these unconscious phenomena--especially since they were such that the patient could become conscious of them through hypnosis or psychotherapy. Freud postulated a mechanism that he called "repression" to explain why the patient is unconscious of them.

In addition to the subliminal and the unconscious, there are more familiar characteristically mental phenomena that do not consist of states of consciousness. When a man falls into a dreamless sleep, he does not lose all his beliefs or abandon all his goals, he does not cease wanting a better world or being artistic or imaginative or lazy, nor does he forget how to do arithmetic or speak French.

A person is not jealous of someone only when thinking of him, nor does a businessman have confidence in the dollar only when concentrating on business. Obviously, these mentalistic characteristics can apply in a dispositional way to people who are not at that moment expressing or exhibiting the disposition.

Furthermore, as Gilbert Ryle has pointed out in great detail, a person may use his mind on many occasions without the feeling of subjective experiences. As he says:

> "When we describe people as exercising qualities of mind, we are not referring to occult episodes of which their overt acts and utterances are effects; we are referring to those overt acts and utterances themselves."

To be responsive to one's surroundings, to act intelligently, deliberately, with wit or good grace, to utilize arithmetic or logic, to be sympathetic or cold-hearted, to drive alertly or absentmindedly--none of these requires the occurrence of subjective experiences or inner states of consciousness, the immediacy of feelings or sensations.

In such activity, there may be nothing going on except performances of a particular kind, and there may be nothing more required except that under further circumstances other performances of a particular kind will be forthcoming. It is, thus, reasonable to conclude that subjective experience is not a necessary condition for the mental.

It would be rash, however, to draw the further conclusion that subjective experiences are in no way involved in whatever is mental. Returning to the case of Leibniz' *petites perceptions* that are not experienced, a person can be conscious of them in various ways, either before getting used to them or when they are alone or when their intensity or his own sensitivity is increased; and Freud's unconscious phenomena can become conscious phenomena under favourable conditions.

The beliefs that an individual is not aware of in sleep are sometimes the objects of his consciousness, as are his moments of laziness and imaginativeness, his knowledge of arithmetic, and his goals. It is dubious that something that has no connection with states of consciousness could qualify as mental.

PSYCHOLOGICAL ASPECTS OF RITES OF PASSAGE

Less scholarly attention has been given to psychological than to social or cultural aspects of rites of passage, in large part because the scholars concerned with such rites in world societies have been principally anthropologists, who lean toward socio-cultural interpretations.

As the foregoing discussion of passage rites in social context illustrates, psychological aspects of rites nevertheless enter strongly if often implicitly into anthropological interpretations as fundamental matters in social solidarity and social disorder.

Emotional ties to kin and other members of society, personal identification with social groups and religious statuses, and commitment to religious ideology and other values are reinforced and sometimes created by rites of passage.

In a realistic sense, the rites serve as blueprints for social relations and religious behaviour, making clear the acceptable ways to act and at the same time pointing up and reinforcing affective relations with other people and with the supernatural.

Familial rites of ancestor worship, for example, are not only reinforcements of familial solidarity but also have psychological value in reinforcing emotional ties among relatives.

Psychological interpretations of passage rites have given greatest emphasis to their value in allaying personal anxiety. Recurrent features of the rites are acts of magic that assure that the outcome of the endeavour will be successful. In the words of the anthropologist Bronislaw Malinowski these acts serve symbolically and psychologically "to bridge over the dangerous gaps in every important pursuit or critical situation" that exist because of man's lack of control of the universe.

By such magical means as miniature boats floated in streams or carried away by the tide, the dead are shown symbolically to go successfully to the other world, and childbirth and successful maturation are similarly depicted magically.

The subjects of rites of passage frequently act out their future roles to the approval of all others. Numerous acts of magic that are not essential to changes in social status may be incorporated in rites of passage and may be seen to give psychological assurance relating to the future life of the individual.

Traditional Japanese practices at childbirth, for example, required that when a girl was born, the placenta be buried in the ground outside the entrance to the dwelling to insure that the girl, when mature, marry in normal fashion and leave the family. When a boy was born, the placenta was buried inside the house to ensure that he remain at home when mature.

The ordeal that a young man or young woman must often undergo during rites of coming-of-age may similarly be seen to provide psychological assurance of success in the new status. Ordeals of this kind are characteristically uncomfortable or frightening, but they are events that any human being ordinarily can endure.

The psychotherapeutic value of passage rites surrounding events in which stress may be acute, such as childbirth, death, and serious illness, is clearly apparent and essentially follows the principles of modern secular psychotherapy.

The subject is made the centre of concentrated attention by many people, is given reassuring evidence of their regard for him and, by means of magic and the intervention of supernatural beings, is assured of a successful outcome. These events are carried out on a high emotional pitch, which gives them added force.

When anxiety is induced by religious beliefs themselves, such as by ideas that if ritual acts are not performed calamitous results will follow, the rites of passage may be said both to create and to allay anxiety.

Where particular social statuses have special honour and prestige, the mere existence of these statuses offers opportunities for gaining psychological satisfaction, and the requirements for gaining these statuses serve to guide behaviour in socially approved channels that offer psychological satisfaction.

Other interpretations of psychological aspects of passage rites have relied upon ideas derived from or inspired by the psychoanalysts. These have sometimes concerned the symbolism involved in the rites and, in anthropological interpretations, have dealt with both Freudian ideas of symbols and the social order.

The psychologist Bruno Bettelheim has interpreted cicatrisation (inducement of scars) of males in rites at coming-of-age as symbolic wounds indicating subconscious male envy of the vagina, the counterpart of Freud's idea of penis envy.

A psychologically oriented anthropologist J.M. Whiting, and others have combined sociological and psychoanalytic theories in attempting to explain why male initiation ceremonies are conducted in some societies and not in others.

Harsh rites, sometimes including genital operations, are held to be correlated with societies in which infant males have long and intimate contact with their mothers, and husbands are prohibited from sexual intercourse with their wives for a period of two years or more.

The long and exclusive relationship between mother and son is assumed to lead to strong emotional dependence upon the mother by the son, which becomes potentially disruptive at the time the son reaches puberty.

The harsh rites are seen to break the bond of dependency and avoid potential social disruption that might otherwise result from discord between son and father at this time.

13. ETHICS AND MORALITY

Many modern writers on ethics share a view of the nature of practical reason derived from Hume. Our reasons for acting morally, they hold, must depend on our desires because reason in action applies only to the best way of achieving what we desire. This view of practical reason virtually precludes any general answer to the question "Why should I be moral?" Until very recently, this question had received less attention in the 20th century than in earlier periods. In the early part of the century, such intuitionists as H.A. Prichard had rejected all attempts to offer extraneous reasons for being moral. Those who understood morality would, they said, see that it carried its own internal reasons for being followed. For those who could not see these reasons, the situation was reminiscent of the story of the emperor's new clothes.

The question fared no better with the emotivists. They defined morality so broadly that anything an individual desires can be considered to be moral. Thus there can be no conflict between morality and self-interest, and if anyone asks "Why should I be moral?" the emotivist response would be to say "Because whatever you most approve of doing is, by definition, your morality."

Here the question is effectively being rejected as senseless, but this reply does nothing to persuade the questioners to act in a benevolent or socially desirable way. It merely tells them that no matter how antisocial their actions may be, they can still be moral as the emotivists define the term.

Other philosophers have put the question to one side, saying that it is a matter for psychologists rather than for philosophers. In earlier periods, of course, psychology was considered a branch of philosophy rather than a separate discipline, but in fact psychologists have also had little to say about the connection between morality and self-interest.

In *Motivation and Personality* (1954) and other works, Abraham H. Maslow developed a psychological theory reminiscent of Shaftesbury in its optimism about the link between personal happiness and moral values, but Maslow's factual evidence was thin. Victor Emil Frankl, a psychotherapist, has written several popular books defending a position essentially similar to that of Joseph Butler on the attainment of happiness. The gist of this view is known as the paradox of hedonism. In *The Will to Meaning* (1969), Frankl states that those who aim directly at happiness do not find it; those whose lives have meaning or purpose apart from their own happiness find happiness as well.

The U.S. philosopher Thomas Nagel has taken a different approach to the question of how we may be motivated to act altruistically. Nagel challenges the assumption that Hume was right about reason being subordinate to desires. In *The Possibility of Altruism* (1969), Nagel sought to show that if reason must always be based on desire, even our normal idea of prudence (that we should give the same weight to our future pains and pleasures as we give to our present ones) becomes incoherent.

Once we accept the rationality of prudence, however, Nagel argued that a very similar line of argument can lead us to accept the rationality of altruism--*i.e.*, the idea that the pains and pleasures of another individual are just as much a reason for one to act as are one's own pains and pleasures. This means that reason alone is capable of motivating moral action; hence, it is unnecessary to appeal to self-interest or benevolent feelings. Though not an intuitionist in the ordinary sense, Nagel has effectively reopened the 18th-century debate between the moral sense school and the intuitionists who believed that reason alone can play a role in action.

The most influential work in ethics by a U.S. philosopher since the early 1960s, John Rawls's *Theory of Justice* (1971), is for the most part centred on normative ethics, and so will be discussed in the next section; it has, however, had some impact in metaethics as well. To argue for his principles of justice, Rawls uses the idea of a hypothetical contract, in which the contracting parties are behind a "veil of ignorance" that prevents them from knowing any particular details about their own attributes. Thus one cannot try to benefit oneself by choosing principles of justice that favour the wealthy, the intelligent, males, or whites.

The effect of this requirement is in many ways similar to Hare's idea of universalizability, but Rawls claims that it avoids, as the former does not, the trap of grouping together the interests of different individuals as if they all belonged to one person. Accordingly, the old social contract model that had largely been neglected since the time of Rousseau has had a new wave of popularity as a form of argument in ethics.

The other aspect of Rawls's thought to have metaethical significance is his so-called method of reflective equilibrium--the idea that a sound moral theory is one that matches reflective moral judgments. In *A Theory of Justice* Rawls uses this method to justify tinkering with the original model of the hypothetical contract until it produces results that are not too much at odds with ordinary ideas of justice. To his critics, this represents a re-emergence of a conservative form of intuitionism, for it means that new moral theories are tested against ordinary moral intuitions.

If a theory fails to match enough of these, it will be rejected no matter how strong its own foundations may be. In Rawls's defence it may be said that it is only our "reflective moral judgments" that serve as the testing ground--our ordinary moral intuitions may be rejected, perhaps simply because they are contrary to a well-grounded theory. If such be the case, the charge of conservatism may be misplaced, but in the process the notion of some independent standard by which the moral theory may be tested has been weakened, perhaps so far as to become virtually meaningless.

Perhaps the most impressive work of metaethics published in the United States in recent years is R.B. Brandt's *Theory of the Good and the Right* (1979). Brandt returns to something like the naturalism of Ralph Barton Perry but with a distinctive late 20th-century American twist. He spends little time on the concept of good, believing that everything capable of being expressed by this word can be more clearly stated in terms of rational desires. To explicate this notion of a rational desire, Brandt appeals to cognitive psychotherapy.

An ideal process of cognitive psychotherapy would eliminate many desires: those based on false beliefs, those which one has only because one is ignoring the feelings or desires that are likely to be expressed in the future, the desires or aversions that are artificially caused by others, desires that are based on early deprivation, and so on. The desires that an individual would still have, undiminished in strength after going through this process, and are what Brandt is prepared to call rational desires.

In contrast to his view of the term good, Brandt does think that the notions of morally right and morally wrong are useful. He suggests that, in calling an action morally wrong, we should mean that it would be prohibited by any moral code that all fully rational people would support for the society in which they are to live. (Brandt then argues that fully rational people would support that moral code which would maximize happiness, but the justification of this claim is a task for normative ethics, not metaethics.)

Brandt's final chapter is an indication of the revival of interest in the question, as he phrases it, "Is it always rational to act morally?" His answer, echoing Shaftesbury in modern guise, is that such desires as benevolence would survive cognitive psychotherapy, and so a rational person would be benevolent. A rational person would also have other moral motives, including an aversion to dishonesty.

These motives will occasionally conflict with self-interested desires, and there can be no guarantee that the moral motives will be the stronger. If they are not, and in spite of the fact that a rational person would support a code favouring honesty, Brandt is unable to say that it would be irrational to follow self-interest rather than morality. A fully rational person might support a certain kind of moral code and yet not act in accordance with it on every occasion.

As the century drew to a close, the issues that divided Plato and the Sophists are still dividing moral philosophers. Ironically, the one position that now has few defenders is Plato's view that "good" refers to an idea or property having an objective existence quite apart from anyone's attitudes or desires--on this point the Sophists appear to have won out at last. Yet, this still leaves ample room for disagreement about the extent to which reason can bring about agreed decisions on what we ought to do. There also remains the dispute about whether it is proper to refer to moral judgments as true and false.

On the other central question of metaethics, the relationship between morality and self-interest, a complete reconciliation of the two continues to prove--at least for those not prepared to appeal to a belief in reward and punishment in another life--as elusive as it did for Sidgwick at the end of the 19th century.

14. HEALTH AND DISEASE

REPRESSED MEMORY THERAPY (RMT)

In 1994 the mental health profession found itself deeply divided over an approach to psychotherapy known as *"repressed memory therapy,"* or RMT. RMT relies on so-called memory-recovery techniques to help a patient "remember" or "recover" episodes, usually of sexual abuse, from childhood--episodes that presumably have been "forgotten." The abuse is assumed to be an underlying cause of the patient's current symptoms.

Repressed memory therapy is based on the theory that in order to cope with the trauma of being abused, the victim employed a psychological defence known as dissociation. Dissociation involves "splitting off" awareness so that the conscious mind is "elsewhere" when the abuse takes place. The result is repression, a self-protective memory loss, or amnesia.

Despite the fact that the painful experiences are consciously forgotten, the repressed material can still cause severe symptoms; often these symptoms have no clear cause. The therapist's role in RMT is to help the patient recover the memories. Presumably, once the memories have been brought into awareness, the survivor's present problems can be effectively treated.

The professional controversy over RMT centres on questions such as:

- Why and under what conditions does an individual repress traumatic memories?

- Can one completely forget repeated episodes of childhood sexual abuse?

- Might the memories that are recovered have been "manufactured" to accommodate the expectations and suggestions of the therapist or to account for otherwise puzzling symptoms?

- Is it possible for a therapist to lead a patient to believe that he or she was sexually abused when no such event actually occurred?

- And perhaps most important, how should therapists conduct themselves when the answers to these questions remain unclear?

Mental health professionals generally agree that sexual abuse of children has been and is a widespread problem. They recognize that historically survivors of such abuse have been discounted owing to a "cultural denial" that has minimized both the scope and the seriousness of the offence. They know that creating a climate in which survivors can come forward, disclose what happened to them, and be believed is vital to their eventual recovery. Beyond this, however, opinions diverge.

On one side are those who believe that dissociation and repression are common responses to sexual abuse in childhood and that victims generally can be readily identifiable from a known list of symptoms. In their view, treatment should first lift the veil of repression through techniques such as hypnosis or "guided imagery." The therapist then must help the patient deal with the painful memories.

Proponents of RMT are concerned that any disbelief in these assumptions on the part of therapists makes it more difficult for sexual abuse survivors to disclose their problems and easier for perpetrators to evade responsibility for their terrible deeds. They reject the notion that detailed traumatic memories can arise merely on the basis of suggestion, contending that such memories need to be acknowledged as true before treatment can succeed.

On the other side of the controversy are those who view dissociation and repression as uncommon responses. They have grave doubts that anyone can suffer repeated trauma over a long period of time and repress all the memories, only to recover them many years or even many decades later under the influence of therapy (or some other suggestive source, such as a book or talk show). RMT opponents do not believe that victims can be identified on the basis of a "symptom checklist."

Furthermore, they hold that it is unsound to hypothesize a history of abuse on the basis of symptoms that might be explained by other means. They recognize that some people may be particularly vulnerable in certain contexts--for example, psychotherapy--and thus may accept "evidence" that has no basis in fact.

When therapists conclude that a patient has been sexually abused, they may lead that patient, intentionally or unintentionally, to reach the same conclusion. Consequently, appropriate treatment for that patient is delayed or even prevented. And finally, RMT opponents are concerned that innocent people will be falsely accused of perpetrating abuse, and their lives and families will be destroyed as a result.

Professionals on both sides of this controversy argue their points vehemently and intelligently, and both groups are motivated by a desire to help "victims" or potential victims.

In order to accommodate the theory of repressed memories, many U.S. states have passed laws allowing a "delayed discovery" to serve as the basis for civil suits. Otherwise, the statute of limitations in these cases would have expired. By the end of 1994, about 500 legal actions had been initiated--most often by a daughter against an allegedly abusive parent. Within the past two years, several cases have been highly publicized, bringing the repressed memory therapy controversy before the public.

One such case involved Gary Ramona, a successful northern California winery executive, whose daughter Holly had sought psychotherapy for depression and bulimia (a severe eating disorder) when she was a college student. During her treatment in 1989-90, she began to recall scenes of sexual abuse from her childhood and came to believe they involved her father.

In the process of therapy, she was given sodium amytal, the so-called truth serum, in an attempt to validate her conclusion that her father had abused her. While under the influence of the drug, she recounted specific instances of abuse by her father.

When Ramona was publicly accused of child sexual abuse, he vehemently denied it. Nonetheless, he lost his job, his wife left him, and his two other daughters cut off all contact with him. His reputation, family, and career thus ruined, Ramona filed an $8 million malpractice suit against Holly's former therapists and the medical centre at which they worked.

He claimed they had planted inaccurate and damaging information in his daughter's mind and had used questionable techniques to do so. On May 13, 1994, the jury in the Napa county superior court where the case was tried awarded a judgment of $500,000 to Gary Ramona. The judgment was not based on the truth or falsity of Holly's memories but on how the memories were obtained. The jury believed the therapists had not conducted themselves appropriately in Holly's treatment.

This case was significant because it was the first repressed-memory case in the United States in which a third party was awarded damages. Normally, if a therapist is sued, it is by a patient. Previous "standard-of-care" cases had considered only the therapist's responsibility to his or her patients, not to the patients' relatives. Another interesting aspect of the Ramona case was that the patient, Holly, testified on behalf of her therapists and continued to maintain that the abuse took place even after the court's decision.

In another widely publicized case, 34-year-old Stephen Cook of Philadelphia filed a $10 million suit against Joseph Cardinal Bernardin of

Chicago, claiming Bernardin had sexually abused him nearly two decades earlier when Cook was a seminary student. Cook had also accused another clergyman at the school of having molested him. Cook's memories of abuse by Bernardin came later and were obtained under hypnosis.

The cardinal quickly and convincingly denied the accusations, bringing Cook's credibility into question. Cook then consulted a psychologist to evaluate his "memories" and determine whether he might have been influenced by the hypnotist. Subsequently, Cook acknowledged publicly that his memories were "unreliable," and in early 1994 he withdrew his allegations against Bernardin. Cook's accusation and subsequent retraction raised many doubts in the public's mind about the validity of RMT. (Cook's case against the other priest was later settled out of court.)

In one of the most bizarre cases involving repressed memories, Paul Ingram, a former Washington state deputy sheriff, was accused by one of his daughters of sexual abuse. The charge stemmed from a memory that surfaced at a church retreat, where the subject of sexual abuse was discussed. Another of Ingram's daughters, who was at the same retreat, then claimed she, too, had been sexually abused by her father. Ingram's law-enforcement colleagues encouraged him to confess because that would help him "remember" the acts he must be repressing.

Although the daughters' allegations were improbable--expanding to include many of Ingram's fellow officers as accomplices and to involve satanic rituals and even human sacrifices (all charges for which no evidence was ever found)--the investigators in this case believed the abuses had occurred. They did not think such detailed stories could be entirely untrue.

A deeply religious man, Ingram reasoned that his daughters would never make up such charges; despite having no such memories, he concluded that the accusations must be true. Believing that neither God nor his daughters would lead him to imagine unfounded guilt, Ingram confessed and went to prison. His intense religiosity led him to believe that any image he conjured up in his mind of having perpetrated abuse must have been placed there by God, thereby confirming his "guilt."

Without objective corroborating evidence--such as a photograph or videotape--how can a real memory of child sexual abuse be distinguished from an illusory one? At present, no reliable method exists for distinguishing truth from fiction in RMT cases.

Clearly, the issues are complex, and courts have been asked to rule even in the absence of hard data. In 1990, for example, in the case in California of 30-year-old Eileen Franklin-Lipsker's recovered memories, the jury convicted her father, George Franklin, in the 1969 murder of one of her childhood friends after Franklin-Lipsker's testimony and that of child psychiatrist Lenore Terr convinced the jury beyond a reasonable doubt that Franklin was guilty. He was convicted of first-degree murder and sentenced to life in prison.

As pressure increases from within the field to approach these sensitive cases with extreme caution, undoubtedly more careful research will be conducted. And as legal rulings shape public perception and further define professional responsibilities, the intensity of the RMT controversy is likely to diminish.

In the meantime, therapists must, as always, honour the Hippocratic oath: *"Primum non nocere"* ("Above all do no harm").

15. CONTRIBUTORS TO PSYCHOTHERAPY

If one compares the three principal tendencies in (and original contributors to) psychotherapy (Freud, Jung, and Adler) with regard to the direction in which their central thought leads, one could say:

1. The analytical method of Sigmund Freud looks for the *causae efficientes*, the causes of the later behavioural disturbances.

2. Alfred Adler considers and treats the initial situation with regard to a *causa finalis* and both see in the drives the *causae materiales*.

3. In Carl Gustav Jung's case the term *'synthesis'* is based on his abandonment of the causal thinking of the alternative psychological methods of treatment. Jungian psychotherapy, therefore, is not an analytical procedure in the usual meaning of this term.

Whatever the differences among Freud's, Jung's and Adler's extensive works on the therapeutic methodologies; scientists, artists, thinkers and practitioners accept the great importance of Freud's and Jung's studies for medicine, psychology, anthropology, religion, art, history, literature...

SIGMUND FREUD

More than seventy-two years have passed since the death of Sigmund Freud (1856-1939), the Austrian psychiatrist and medical consultant who invented the use of morphine and proceeded to establish himself as the father of Psycho-analysis. The agora is still in full session for the debating as to whether he was a genius and if his works had a fruitful impact on the treatment of neurosis and society at large.

CONSCIOUS AND SUBCONSCIOUS

Sigmund Freud proposed the idea of *conscious and subconscious* mind. He began psycho-analysis and proposed theories of infantile sexuality and their effects on adult life. His volume of work on The Interpretation of Dreams influenced the world of art, and many authors based their novels on his writings.

At the peak of his career, Sigmund Freud was named as the Copernicus of the Mind. Inspired by Goethe's essay on Nature, he studied medicine in Vienna, but original work in physiology delayed his graduation. He then studied and specialised in neurology, spurred on by physician Breuer, that hysteria can be cured by making a patient recall painful memories under hypnosis, studied under Charcot in Paris and changed over from neurology to psychopathology. To appease his frowning colleagues in Vienna, he published on his return two strictly neurological studies on aphasia and cerebral paralysis, before risking with Breuer, the joint publication of Studien über Hysterie.

HYPNOSIS

Finding hypnosis inadequate, Freud gradually substituted the method of 'free association'. Allowing the patient to ramble on with his or her thoughts when in a state of relaxed consciousness and, interpreting the data, an abundance of childhood and dream recollections. He became convinced, despite his own puritan sensibilities, of the fact of infantile sexuality. This became the basis of his theory and cost him his friendship with Breuer, lost him many patients and isolated him from the always conservative medical profession.

REPRESSION

Thereafter, he worked alone, publishing many papers and books, which included his work on dreams, which showed that dreams, like neuroses, are disguised manifestations of repressed wishes of sexual origin. Repression, which differs profoundly from mere conscious suppression, Freud explained by reference to a vast reservoir of subconscious, irrational mental activity, the *Id,* comprising the crude appetites and impulses, loves and hates, particularly those connected with what he termed the Oedipus complex, the infant's craving for exclusive possession of the parent of the opposite sex.

These impulses, at variance with civilised behaviour, are repressed by the ego, a portion of the id which at an early stage has become differentiated from it. At a later stage, the super-ego (conscience) develops out of the ego, determining what is acceptable to the ego and what must be repressed.

Repressions disappear from consciousness but live on in the id. In sleep or in day-dreaming the ego relaxes its control and the repressed impulses may succeed in pushing themselves into consciousness, but not until the reduced powers of the ego have exercised a censorship, by distorting the unacceptable character of the dream material into something meaningless but acceptable. Psycho-analysis seeks to uncover these repressions of what they are and replace them by acts of judgement.

JUNG AND ADLER

Further works, writings and essays by Freud met with heated, incomprehensive opposition and was not before 1930, when Freud was awarded the Goethe prize, that his efforts no longer aroused active opposition from public bodies. This award was bestowed to Freud after Jung and Adler diverged from the Freudian theory by seeking to remove the central emphasis on sexuality.

BREAKING UP

Adler, who broke with Freud in 1911, developed a psychology of the ego later known as Individual Psychology, and Jung, who followed in 1913, developed a highly complex system of basic human types and the 'collective subconscious, which later on were known as the psychotherapy of Jung. Thereafter, Ernest Jones formed a committee of senior collaborators pledged to uphold the basic Freudian conceptions. Psychoanalysis thenceforth was a creed as well as a science.

INTER-RELATIONSHIP OF SCIENCES

For those of us who studied, practised, applied a variety of methods, and inter-related the sciences have no doubt as to the effect on everyday life and the way of living in Western societies. The terminology established by Sigmund Freud is present in the vocabulary of everyday life and the interpretation of his Eros concept is still predominant in the advertising media professions.

16. JUNG, THE PSYCHOTHERAPIST

Carl Gustav Jung established the term of Psychotherapy as a part of the wider aspect of his psychological and psychiatric studies. In this manner, the Jung Psychotherapeutic works are divisible into a theoretical part, whose principal headings can be described quite generally as:

- **Nature and Structure of the Psyche,**

- **Laws of the Psychic Processes and Forces,**

- **The practical part based on these theories, their application, as therapeutic method in the narrower sense.**

PHILOSOPHICAL DERIVATION

If one would reach a correct comprehension of Jung's *system*, one must first of all accept Jung's standpoint and recognise with him the full reality of the *psychic functions*. This point of view was, remarkable as it may sound, relatively new at his time. For up to a few decades earlier, the *psyche* was not considered as independent and subject to its own laws, but was studied and interpreted through derivation from philosophy, religion or from natural science, so that its true nature could not rightly be discerned.

PSYCHIC EQUALS PHYSICAL

To C G Jung the psychic is no less real than the physical. Though it is not immediately touchable and visible, it is still fully and unambiguously experienceable. Even in the twenty-first century, it is a world in itself – subject to law, structured, and possessed of its special means of expression. All that we know of the world comes to us, as does all knowledge of our own being, through the medium of the *psychic*, which is therefore one of the most important aspects and conditions of experience.

To study it as such was Jung's aim; not however to elevate it as would a mere psychologism to be the sole ground of all knowledge. The psychological, physical, and physico-mathematical standpoints (as well as many others) are interchangeable and can be studied at will according to the problem and the special interests of the enquirer.

PSYCHOLOGICAL ASPECT

Jung took the psychological aspect, leaving the others to persons competent in their fields, drawing however upon his wide acquaintance with psychic reality, so that this theoretical structure is no abstract system created by the speculative intellect but an erection upon the solid ground of experience and resting only on that.

Its two main pillars were:

- The principle of *psychic totality*,

- The principle of *psychic dynamics*.

These two points were elaborated, together with the practical application of *the system*, in Jung's researches and book publications.

PSYCHE, SOUL, OR MIND

By using the term *'psyche'* Jung understood not merely what we usually mean by the word *'soul'* or *'mind'*, but the totality of all psychological processes, both conscious and unconscious. That is something broader than and including the soul, which for him constituted only certain limited complex of *functions*.

According to his definitions, the *psyche* consists of two spheres supplementing one another but opposed in their properties – of *consciousness* and the so called *unconscious*.

EGO

The *ego* has a share of both. The following diagram shows the ego standing between two spheres, which not only supplement but also complement or compensate each other. That is, the dividing line that marks them off from each other in our ego can be displaced in both directions, as is suggested by the arrows and the dotted lines in the figure.

The ego itself is not exclusively *conscious*, but is conceived as a centre of reference for conscious and unconscious *psychic contents* alike. It forms, as a concept embracing the unitary totality of our *psychosomatic* beings. It is naturally only expedient of thought and an abstraction that the ego stands exactly in the middle.

CONSCIOUSNESS

Jung defines consciousness as *"the function or activity which maintains the relation of the psychic contents to the ego"*. The next diagram shows how the *sphere of consciousness* is surrounded by contents lying in the unconscious. Here are those contents which have been put aside (for our consciousness can take only a very few contents at once) but which can be raised again at any time into consciousness; furthermore, those which can be repressed because they can be disagreeable for various reasons - i.e., *"forgotten, repressed, subliminally perceived, thought, and felt matter of any kind."*

This region Jung called it the *'Personal Unconscious'* in order to distinguish it from that of the *'Collective Unconscious'*. For the *collective part of the unconscious* no longer includes contents that are specific for the *individual ego* and result from the personal acquisitions, but such as result *"from the inherited possibility of psychical functioning in general, namely from the inherited brain structure."* This inheritance is common to all humanity, perhaps even to the entire animal world, and forms the basis of every individual psyche.

Further on, Jung maintained that the *unconscious* is older than *consciousness*. He added that it is the primal datum out of which *'ever afresh arises'*. Thus, consciousness is merely built upon the *fundamental psychic activity*, which consists in the functioning of the unconscious.

PSYCHIC FUNCTIONS

By a psychological function Jung understood a *"certain form of activity that remains theoretically the same under varying circumstances and is completely independent of its momentary contents."* The decisive fact is not what one thinks, but that one employs one's *intellectual function* and not one's *intuition* in receiving and working up contents presented from without or within. *Thinking* is that function which seeks to reach an understanding of the world and an adjustment to it by means of an act of thought, or cognition, i.e., of conceptual relations and logical deductions. In contrast thereto, the *feeling function* apprehends the world on the basis of an evaluation by means of the concepts, pleasant or unpleasant, adience or avoidance.

Both functions are characterised as rational because they work with values: thinking evaluates by means of cognitions from the viewpoint 'true/false, feeling by means of emotions from the viewpoint *'agreeable/disagreeable'*. These two fundamental forms of reactions are mutually exclusive as practical determinants of behaviour; the one or the other predominates.

The other two functions, *sensation and intuition*, Jung called the *irrational functions*, since they circumvent the ratio and work not with judgements but with mere perceptions, without evaluation or interpretation. *Sensation* perceives things as they are and not otherwise. It is the sense of reality par excellence, what the French call the *'fonction du rèel'*. *Intuition* perceives likewise, but less through the *conscious apparatus of the senses* than through its capacity for an unconscious *'inner perception'* of the potentialities in things. The *sensation type* will take notice of an historical event in all its details but disregard the psychological context in which it is set; the *intuitive*, on the contrary, will pass over the details carelessly but perceive without difficulty the inner meaning of the occurrence, its possible relations and consequences.

Although man possesses constitutionally all four functions, experience shows that it is always only one of these functions with which he orientates himself and adjusts himself to reality. This function becomes the dominant function for adjustment; it gives the conscious attitude its direction and quality and stands constantly at the disposal of the individual conscious will.

If we wish to give a complete schematic representation of the personality according to Jung's *typological system*, we can think of *introversion-extraversion* as constituting a third axis perpendicular to the cross axes of the four functional types. Referring each of the four functions to both the attitudinal types, we get an eightfold spatial figure. The idea of the quaternity is in fact seldom expressed by the double four, the eight, as well as by the four itself.

The four functional types, based on the predominance of the one or the other function in the individual are valid in this form only theoretically. In real life they almost never occur pure but more or less as mixed types.

17. LOGIC, PHILOSOPHY OF PSYCHOTHERAPY

Although the "laws of thought" studied in logic are not the empirical generalisations of a psychotherapist, they can serve as a conceptual framework for psychotherapeutic theorising. Probably the best known recent example of such theorising is the large-scale attempt made in the mid-20th century by Jean Piaget, a Swiss psychologist, to characterise the developmental stages of a child's thought by reference to the logical structures that he can master.

Elsewhere in psychotherapy, logic is employed mostly as an ingredient of various models using mathematical ideas or ideas drawn from such areas as automata or information theory. Large-scale direct uses are rare, however, partly because of the problems mentioned above in the section on logic and information.

PSYCHOTHERAPY/SCIENCE

One has to start with one simple question; is psychotherapy a science? In all references and concepts, modern psychotherapists claim that their profession is a scientific one, no matter which branch they follow, and in dealing with living beings their approach is based on science. Psychotherapy and its many schools of ideas, together with all the conceptual comprehension of behaviour, it still comes under the auspices of philosophy and philology.

An individual develops into a 'socialised' human being through interaction with others. A man's values, his language, his ways of behaving are all learned from other people, though certain limits to his development are set by hereditary and innate factors.

Therefore, when a human being needs to talk to someone, or needs some help in coping with life he/she will interact with those close and if not enough response is shown will seek assistance from a professional; psychologist, counsellor, psychotherapist, mental nurse. None of these professionals is a scientist.

MEDICAL DOCTOR

The nearest to a scientist is his/her family doctor who will listen and use a scientific method based the medical training the doctor received. If time does not permit then the medical doctor should refer the case to a psychologist, counsellor, psychotherapist, or a mental nurse. Thereafter, the case ought to be supervised by the doctor, and in return expect a report on the case.

Prescribing medication appropriate to the case comes under the title of science. The scientific modern pharmaceutical drugs have proven to be beneficial to patients. The latest development of medicines and the cure of many types of disorders are based on the research of genetics. Genetics and the cataloguing of genomes are, therefore, scientific studies.

Psychotherapy (and all its branches, whether accepted as a scientific subject or a philosophical sub-division) is a relatively enduring organisation of interrelated beliefs that describe, evaluate, and advocate action with respect to a human being, another animal, an object or situation, with each belief having cognitive, and behavioural components.

END

18. BIBLIOGRAPHY

PUBLICATIONS BY ANDREAS SOFRONIOU USED AS BIBLIOGRAPHY FOR THIS BOOK

PHILOSOPHY
1. MORAL PHILOSOPHY, FROM SOCRATES TO THE 21ST AEON, ISBN: 978-1-4457-4618-0
2. MORAL PHILOSOPHY, FROM HIPPOCRATES TO THE 21ST AEON, ISBN: 978-1-84753-463-7
3. THERAPEUTIC PHILOSOPHY FOR THE INDIVIDUAL AND THE STATE, ISBN: 978-1-4092-7586-2
4. PHILOSOPHIC COUNSELLING FOR PEOPLE AND THEIR GOVERNMENTS, ISBN: 978-1-4092-7400-1
5. MORAL PHILOSOPHY, THE ETHICAL APPROACH THROUGH THE AGES, ISBN: 978-1-4092-7703-3
6. MORAL PHILOSOPHY, ISBN: 978-1-4478-5037-3
EDUCATION
7. 2011 POLITICS, ORGANISATIONS, PSYCHOANALYSIS, POETRY, ISBN: 978-1-4467-2741-6
8. PLATO'S EPISTEMOLOGY, ISBN: 978-1-4716-6584-4
9. ARISTOTLE'S AETIOLOGY, ISBN: 978-1-4716-7861-5
10. MARXISM, SOCIALISM & COMMUNISM, ISBN: 978-1-4716-8236-0
11. MACHIAVELLI'S POLITICS & RELEVANT PHILOSOPHICAL CONCEPTS, ISBN: 978-1-4716-8629-0
12. BRITISH PHILOSOPHERS, 16TH TO 18TH CENTURY, ISBN: 978-1-4717-1072-8
13. ROUSSEAU ON WILL AND MORALITY, ISBN: 978-1-4717-1070-4
14. HEGEL ON IDEALISM, KNOWLEDGE & REALITY, ISBN: 978-1-4717-0954-8
15. PHILOLOGY, CONCEPT OF EUROPEAN LITERATURE, ISBN: 978-1-291-49148-7
16. THREE MILLENNIA OF HELLENIC PHILOLOGY, ISBN:
MEDICINE
17. MEDICAL ETHICS THROUGH THE AGES, ISBN: 978-1-4092- 7468-1
18. MEDICAL ETHICS, FROM HIPPOCRATES TO THE 21ST CENTURY, ISBN: 978-1-4457-1203-1
19. THE MISINTERPRETATION OF SIGMUND FREUD, ISBN: 978-1-4467-1659-5
20. JUNG'S PSYCHOTHERAPY: THE PSYCHOLOGICAL & MYTHOLOGICAL METHODS, ISBN: 978-1-4477-4740-6
21. FREUDIAN ANALYSIS & JUNGIAN SYNTHESIS, ISBN: 978-1-4477-5996-6
22. PSYCHOLOGY FROM CONCEPTION TO SENILITY, ISBN: 978-1-4092-7218-2
23. PSYCHOTHERAPY, CONCEPTS OF TREATMENT, ISBN: 978-1-291-50178-0
PSYCHOLOGY
24. PSYCHOLOGY, CONCEPTS OF BEHAVIOUR, ISBN: 978-1-291-47573-9
25. PSYCHOLOGY OF CHILD CULTURE, ISBN: 978-1-4092-7619-7
26. JOYFUL PARENTING, ISBN: 0 9527956 1 2
27. THE GUIDE TO A JOYFUL PARENTING, ISBN: 978-1-4457-1448-6
28. PHILOSOPHY FOR HUMAN BEHAVIOUR, ISBN: 978-1-291-12707-2